W9-CAG-789

Chet Raymo, a professor of physics and astronomy at Stonehill College in Massachusetts, is also the author of *The Crust of Our Earth* (Prentice-Hall, 1983). Both of Dr. Raymo's previous books were selections of the Book-of-the-Month Club.

PHalarope Books

PHalarope books are designed specifically for the amateur naturalist. These volumes represent excellence in natural history publishing. Each book in the PHalarope series is based on a nature course or program at the college or adult education level or is sponsored by a museum or nature center. Each PHalarope book reflects the author's teaching ability as well as writing ability. Among the books:

The Amateur Naturalist's Handbook
Vinson Brown

The Amateur Naturalist's Diary
Vinson Brown

The Art of Field Sketching
Clare Walker Leslie

The Art of Painting Animals: A Beginning Artist's Guide to the Portrayal of Domestic Animals, Wildlife, and Birds
Fredric Sweney

At the Sea's Edge: An Introduction to Coastal Oceanography for the Amateur Naturalist
William T. Fox/illustrated by Clare Walker Leslie

Biography of a Planet: Geology, Astronomy, and the Evolution of Life on Earth
Chet Raymo

Botany in the Field: An Introduction to Plant Communities for the Amateur Naturalist
Jane Scott

A Complete Manual of Amateur Astronomy: Tools and Techniques for Astronomical Observations
P. Clay Sherrod with Thomas L. Koed

The Crust of Our Earth: An Armchair Traveler's Guide to the New Geology
Chet Raymo

The Curious Naturalist
John Mitchell and the Massachusetts Audubon Society

Discover the Invisible: A Naturalist's Guide to Using the Microscope
Eric V. Gravé

Exploring Tropical Isles and Seas: An Introduction for the Traveler and Amateur Naturalist
Frederic Martini

A Field Guide to the Familiar: Learning to Observe the Natural World
Gale Lawrence

A Field Guide to Personal Computers for Bird Watchers and Other Naturalists
Edward M. Mair

A Field Manual for the Amateur Geologist: Tools and Activities for Exploring Our Planet
Alan M. Cvancara

The Fossil Collector's Handbook: A Paleontology Field Guide
James Reid Macdonald

Nature in the Northwest: An Introduction to the Natural History and Ecology of the Northwestern United States from the Rockies to the Pacific
Susan Schwartz/photographs by Bob and Ira Spring

Nature Drawing: A Tool for Learning
Clare Walker Leslie

Nature Photography: A Guide to Better Outdoor Pictures
Stan Osolinski

Nature with Children of All Ages: Activities and Adventures for Exploring, Learning, and Enjoying the World Around Us
Edith A. Sisson, the Massachusetts Audubon Society

Outdoor Education: A Manual for Teaching in Nature's Classroom
Michael Link, director, Northwoods Audubon Center, Minnesota

The Plant Observer's Guidebook: A Field Botany Manual for the Amateur Naturalist
Charles E. Roth

The Seaside Naturalist: A Guide to Nature Study at the Seashore
Deborah A. Coulombe

Sharks: An Introduction for the Amateur Naturalist
Sanford A. Moss

Suburban Wildflowers: An Introduction to the Common Wildflowers of Your Back Yard and Local Park
Richard Headstrom

Suburban Wildlife: An Introduction to the Common Animals of Your Back Yard and Local Park
Richard Headstrom

Thoreau's Method: A Handbook for Nature Study
David Pepi

365 Starry Nights: An Introduction to Astronomy for Every Night of the Year
Chet Raymo

Trees: An Introduction to Trees and Forest Ecology for the Amateur Naturalist
Laurence C. Walker

The Wildlife Observer's Guidebook
Charles E. Roth, Massachusetts Audubon Society

Wood Notes: A Companion and Guide for Birdwatchers
Richard H. Wood

Biography of a Planet

Geology, Astronomy, and the Evolution of Life on Earth

CHET RAYMO

A SPECTRUM BOOK   Prentice-Hall, Inc., Englewood Cliffs, New Jersey 07632

Library of Congress Cataloging in Publication Data

Raymo, Chet.
    Biography of a planet.

    (PHalarope books)
    "A Spectrum Book."
    1. Life—origin.   2. Life (Biology)   3. Evolution.
4. Geology.   5. Astronomy.   I. Title.
QH325.R39 1984   577   84-4810
ISBN 0-13-078221-1
ISBN 0-13-078213-0 (Pbk.)

© 1984 by Prentice-Hall, Inc., Englewood Cliffs, N.J. 07632

A SPECTRUM BOOK

All rights reserved. No part of this book
may be reproduced in any form or by any means
without permission in writing from the publisher.

10 9 8 7 6 5 4 3 2 1

Printed in the United States of America

This book is available at a special discount when ordered
in bulk quantities. Contact Prentice-Hall, Inc., General
Publishing Division, Special Sales, Englewood Cliffs, N. J. 07632.

Editorial/production supervision: Joe O'Donnell Jr.
Cover design: Hal Siegel
Cover photograph courtesy of NASA
Manufacturing buyer: Doreen Cavallo

ISBN 0-13-078221-1

ISBN 0-13-078213-0 {PBK.}

Prentice-Hall International, Inc., *London*
Prentice-Hall of Australia Pty. Limited, *Sydney*
Prentice-Hall Canada Inc., *Toronto*
Prentice-Hall of India Private Limited, *New Delhi*
Prentice-Hall of Japan, Inc., *Tokyo*
Prentice-Hall of Southeast Asia Pte. Ltd., *Singapore*
Whitehall Books Limited, *Wellington, New Zealand*
Editora Prentice-Hall do Brasil Ltda., *Rio de Janeiro*

Contents

**To my mother, Margaret,
for the words,
and my father, Chester,
for the pictures.**

I would like to thank my sons Dan and Tom and my daughter Maureen for help in preparing the illustrations and text. My wife, Maureen, infused the book with much good sense. Maurice Sheehy and Philip Naughton shared walks of discovery. Mary Kennan, editor in the General Publishing Division at Prentice-Hall, and Joe O'Donnell Jr. and Maria Carella, also in GPD, brought the book into the world with grace and skill.

Introduction

"History is subject to geology," wrote Will and Ariel Durant. "Every day the sea encroaches somewhere upon the land, or the land upon the sea; cities disappear under the water, and sunken cathedrals ring their melancholy bells. Mountains rise and fall in the rhythm of emergence and erosion; rivers swell and flood, or dry up, or change their course; valleys become deserts, and isthmuses become straits. To the geologic eye all the surface of the earth is a fluid form, and a man moves upon it as insecurely as Peter walking on the waves to Christ."

The Durants addressed themselves to the thousands of years of human history. We shall take a longer view, across four billion years of the history of life on Earth. But our conclusion will be the same: History is subject to geology.

The origin of life on Earth was the result of certain favorable circumstances that just happened to exist on the third planet of a yellow star in a suburb of the Milky Way Galaxy. Here—perhaps uniquely—matter rose up and embraced the wind, walked upon the waters, pondered the stars, spoke in tongues.

Life was the child of the stars, an offshoot of cosmic evolution. The story of life is the story of matter and energy. It is the story of forces that reach across light-years, binding stars to galaxies and planets to stars, holding the universe together, containing, squeezing, turning the screws, pressing new things into existence. It is the story of beginnings and endings, of stars born in dark clouds of dust and of stars that die with spasms of light that illuminate galaxies. It is the story of rock and iron, of things that endure across eons, and it is the story of gossamer substances that cling to the planet like dew to the rose, exhalations of the Earth so fragile that the shrug of a star can blow them away.

Once life got its grip on the planet, it did not let go. The Earth was a relatively secure container for life, and the Sun was a relatively steady star. But throughout its history life has been buffeted by huge shifts of the Earth's crust, disturbing variations of climate, rising and falling sea levels, collapsing magnetic fields, supernovas, cosmic rays, and asteroid bombardments. Volcanoes, earthquakes, ice ages, the filling and evaporation of great sea basins—all added an element of uncertainty to life's grasp on the planet.

This book traces the evolution of life on Earth within the physical context, from stardust beginnings to human flight back to the stars. In recent years there have been remarkable breakthroughs in the scientific reconstruction of the geological and astronomial setting for life. Almost every weekly issue of *Science and Nature*, for example, contains reports that add new details to the picture. What has emerged is a view of life on Earth that differs in several important respects from the orthodoxy of yesterday.

The older evolutionary theory stressed constant environmental pressures for change and the incremental response of competing organisms to those pressures. The serious business of evolution was at the level of the gene; patient, invisible, inexorable, slow. Current thinking about evolution accentuates the tendency of organisms to endure in the face of pressure, to maintain a status quo, until moments of extreme stress when bets are called off and life must leap a chasm and adapt quickly or face extinction. The new story of life is a story of equilibrium punctuated by catastrophe.

Today we marvel at life's ability to regulate its own environment, to behave as a kind of thermostat on climate, water temperature, and chemical environment. The newer theories see the Earth whole and life as one component of a complex system linked by feedback loops and mutually regulating clocks and valves. Life is no passive plaything of the environment; life *is* environment.

On such a planet, human intervention in the balance of the equation can be benevolent or catastrophic. We have the technology to short-circuit the thermostats, to break the feedback loops, to reset the clocks, to clog the valves. Technology may be the most serious crisis life has faced in four billion years of evolution.

For the first time in the history of the Earth, the actions of a single species can significantly affect the course of evolution. We need only reflect on the possible consequences of a global nuclear war to grasp the power humans hold over the planet. Many of the chapters of this book begin with a personal anecdote: the view of an island across dark water, a walk to a mossy outcrop on a wooded hill, a Valentine's Day conjunction of planets in the western sky. I do not

apologize. It is when we insist on seeing the story of the evolving Earth as "cold science"—sealed in textbooks, remote from personal experience—that we stand in the greatest danger of losing it all.

Every pebble has a story to tell of the evolution of the Earth. Every blade of grass is a poem of the past. Our own bodies are museums of our history; our cells are the scrapbooks of our microbial predecessors; we breathe the exhalations of bacteria that swam in ancient seas. The story of the Earth is all around us waiting to be read.

So come with me to the forge of atoms in the hearts of stars; watch with me as the mountains rise and fall; walk with me between continent-spanning walls of ice. Consider with me the fate of the methane-respiring bacterium, the thunder-footed dinosaur, and the wily opossum. Know with me that the story of the Earth is *our* story. Care with me for the blue-white planet that is our home.

Biography
of a Planet

The Orion
Connection

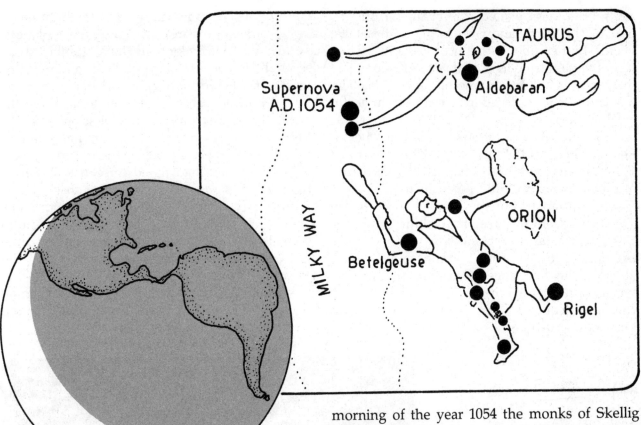

We are made of the ash of stars. Stars burn hydrogen, the most common element in the universe. The ash is carbon, oxygen, nitrogen, and other elements of life.

Skellig Michael is a pyramid of rock that juts like a fractured bone from the Atlantic Ocean near the coast of Ireland. Like many high crags on Europe's shore, the place is dedicated to the Archangel Michael. It is a steep, unsoftened place, more like an eagle's aerie than a place fit for men.

From the 6th or 7th century onward, a community of monks inhabited this beautiful, desolate isle. Their beehive-shaped stone cells are still perched on precipitous cliffs 500 feet above the sea. Here for centuries they lived and worked and prayed and occasionally fended off Viking raids. And here they watched the sky.

I have sketched a view from Skellig Michael on a morning in early September of A.D. 1054. The stars burn like diamonds in a sky that is just beginning to show the light of dawn. The view is to the southeast. Across a stretch of coastal water, rocky headlands separate the dark sky from the answering darkness of the sea.

I would like to imagine that on that particular morning of the year 1054 the monks of Skellig Michael turned from prayer to gape at an apparition—divine signal? dire portent?—that had for some weeks disturbed their familiar sky.

If you know the stars, you will recognize the Milky Way plunging from the zenith to the horizon, a luminous river shoaled with dark reefs. And you will recognize the prominent constellations of Orion, the Hunter, and Taurus, the Bull. Among the many bright stars in this morning's sky, orange Aldebaran, ruddy Betelgeuse, and blue-white Rigel burn with a wintry fierceness.

If you are not familiar with the night sky, the sketch on this page will help you to identify the stars and constellations. The sketch also shows the planet Earth poised in space at that moment in A.D. 1054 when the monks on Skellig Rock looked out into the night.

But wait! The experienced starwatcher will know, as the monks knew, that something is amiss. There, at the tip of the Bull's lower horn, is an unexpected star. What's more, this interloper shines more brightly than anything else in the morning sky, a dozen times more luminous than gorgeous Rigel.

This new star in the horn of the Bull had been visible since early July when it appeared unheralded, a sudden terrible beacon. For a time it outshone even Venus at its brightest. During July and August the star slowly faded. But as the star's brightness waned, the changing orientation of the Earth and Sun brought the stellar visitor higher above the horizon and farther from the obscuring light of dawn. By September the star was still bright enough to be conspicuous and was well placed for viewing from northern latitudes.

Exactly when the monks on Skellig Michael first took note of the new star we do not know. They left no record of the event. The appearance of the star was carefully noted in the annals of China and Japan, but the keepers of European chronicles seem to have shown little interest in this spectacular celestial display. Perhaps the prevalence of Aristotle's philosophy, which stressed the unchangeableness of the celestial sphere, prevented Europeans from recognizing the significance of the new star in Taurus. Or—more simply—horizon haze, the light of dawn, a spell of bad weather, and the location of the new star in a part of the sky where planets are known to wander may have combined to lessen the impact of the event. Within 22 months the star had faded from sight and from memory.

It is with objects of the same nature as the new star in Taurus that the biography of the Earth must begin. Other new stars had flared in the Milky Way Galaxy before the intruder of A.D. 1054. New stars had flared unseen long before the Earth was born from the dust of space. Those luminous transients were not, in fact, "new stars." They were dying stars, stars that die in spasms of light that can illuminate whole galaxies. They were the factories in which the atoms of the Earth were forged.

Walt Whitman was literally correct when he sang "I believe a leaf of grass is no less than the journey-work of the stars." Leaves of grass, like

The Great Nebula
in Orion

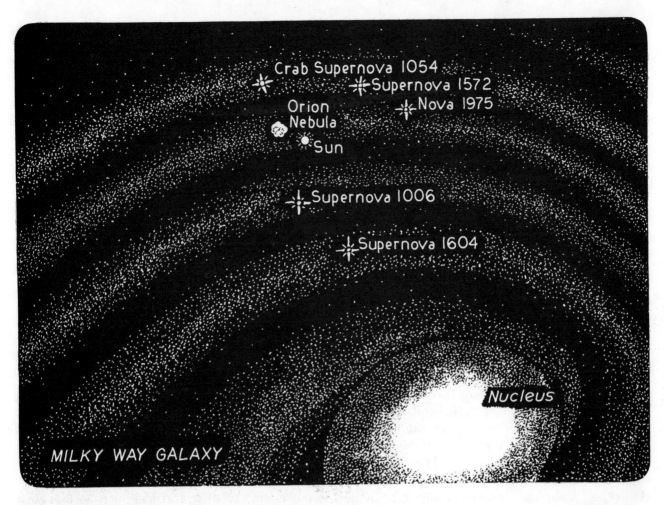

Crab Supernova 1054
Supernova 1572
Orion
Nebula
Nova 1975
Sun
Supernova 1006
Supernova 1604
Nucleus
MILKY WAY GALAXY

all living things on Earth and the planet itself, are composed of atoms. They are composed of hydrogen, the lightest and most common of all atoms, with a single nuclear proton. And they are composed of heavier atoms like oxygen, carbon, and nitrogen, with nuclei containing many protons and neutrons. According to modern theories for the creation of the elements, the heavier elements are "cooked up" in the hot interiors of stars. As stars burn they fuse heavy elements from light ones, making carbon and oxygen and iron (for example) from the nuclei of hydrogen and helium. These heavy elements are then flung into space in the spectacular convulsions of dying stars called novas and supernovas. The new star that blazed out in Taurus in A.D. 1054 was a supernova.

But before we follow the fate of dying stars, let's go back to the sky above Skellig Michael. Drop your gaze to the region of Orion's belt, and particularly to the three stars that are traditionally the hunter's sword. If the night is clear you will be able to tell that the middle star of the sword lacks the sharp clarity of the other stars. Binoculars or a small telescope will reveal that the "star" is not a star at all but a swirling drapery of luminous gas. The gas is heated to fluorescence by the radiation of hot young stars nested in the heart of the cloud.

This is the Great Nebula in Orion, a gas cloud of cosmic dimensions, more than 20,000 times wider than the entire Solar System. Radio telescopes reveal that the Great Nebula is just one bright corner of an immense dark cloud that fills the body of Orion. In that dark ocean of gas brilliant blue-white stars burn like island jewels.

The nebula is 1600 light-years from the Earth, far beyond the boundaries of the Solar System, far beyond our neighboring stars. (Our nearest stellar neighbor, Alpha Centauri, is 4.3 light years distant.) The distance to the Orion complex of gas and stars is 10 quadrillion miles—1 followed by 16 zeros. The mind must struggle to comprehend such distances. To reach the Orion nebula the Voyager space craft would have to travel for 20 million years. And yet, on the still grander scale of the Milky Way Galaxy, the Orion complex is relatively nearby.

Our Sun is just one of several hundred billion stars that swim in the flat whirlpool of the

Milky Way Galaxy. The Sun lies near the inside edge of one of the Galaxy's spiral arms. The Orion gas cloud lies in the same spiral arm near its lower edge. Many other rich clouds of gas and dust decorate the starry arms of the Galaxy. It is in these clouds that astronomers believe stars and planets are born, pulled together by gravity from the stuff of the clouds.

The Orion complex of gas and dust is the nearest of these star-forming regions to the Sun, and that is why the hot young stars of Orion burn so brilliantly in our night sky. It is a stellar nursery that offers astronomers an unparalleled opportunity to study the process of star formation. In the Orion cloud alone there is sufficient gas to make a hundred thousand Suns.

Almost all of the atoms in these great interstellar gas clouds are hydrogen and helium, the lightest of the elements and the most common elements in the Universe. The hydrogen and most of the helium is primordial. It was created, according to the best current theories, 15 billion years ago with the Universe itself, in a "Big Bang" explosion from a singular and infinitely dense seed of energy. Using well-known laws of physics, astronomers can calculate the conditions that existed in the earliest moments of the Universe, and reconstruct the creation of matter from energy which occurred at that time. The calculations rule out the creation of elements heavier than helium, with the possible exception of small amounts of lithium and beryllium.

If the present universe contained only hydrogen and helium, as it did in the early days, then there could be no planet Earth with an iron core and a rocky crust. Nor would the planet's rocky surface be softened by leaves of grass swept by wind and rain and alive with the chemistry of oxygen and carbon.

In the dark clouds of Orion, astronomers can map the presence of carbon, oxygen, nitrogen, and tiny grains of dust. We are ourselves made of such stuff. If only hydrogen and helium were created with the Universe, then where did these heavy elements come from?

The secret of the heavy elements lies in the burning that goes on at a star's core. Astronomers have a fairly confident theoretical picture of how stars burn. The terrible reality of hydrogen bombs confirms the correctness of their calculations. The likely truth is this: Stars produce their prodigious energy by fusing heavy elements from light ones.

A simplified version of the nuclear reactions that take place at a star's core is sketched on the facing page. In the diagram, the dark circles represent protons and the light circles are neutrons. The assemblies of these particles are the nuclei of atoms.

Like the clouds of gas out of which they are formed, the stars are mostly hydrogen. The nucleus of a hydrogen atom is a single proton, a positively charged particle that is one of the fundamental building blocks of nature. The protons are the starting point for the reactions that take place at the cores of stars.

Under ordinary conditions, two positive electrical charges will repel each other. Under ordinary conditions, then, two protons are mutually repulsive. But if the temperature of a "soup" of protons is high enough (8 million degrees Centigrade), the protons will overcome their mutual repulsion and stick together. The new larger units are held together by an attractive nuclear force. In the process of their sticking together, energy is released!

Temperatures high enough to fuse protons occur naturally in the dense interiors of stars, as a result of the squeezing pressure of the overlying matter. Stars are pulled together by gravity from clouds of dust and gas such as the great dark clouds in Orion. When the temperature at the core of a collapsing cloud reaches 8 million degrees, fusion begins. The pressure of the energy released by fusion balances the squeeze of gravity. The cloud stops its collapse, and a star is born.

As a star burns, hydrogen is turned into helium by the steps shown in the top half of the diagram. This conversion of hydrogen to helium, astronomers believe, is presently occurring at the center of the Sun. This is the process that generates the Sun's heat and light.

If you look carefully at the diagram, you will see that the net effect of the process is to convert four protons and two electrons into the nucleus of a helium atom. Separately, the four protons and two electrons have a total mass that is slightly greater than the mass of the helium nucleus formed by their combination. What happened to the tiny bit of mass that is lost in the process of fusion? The answer: The missing mass is converted into pure energy according to Einstein's famous formula $E=mc^2$. Multiply the missing mass by the speed of light squared, and you get energy.

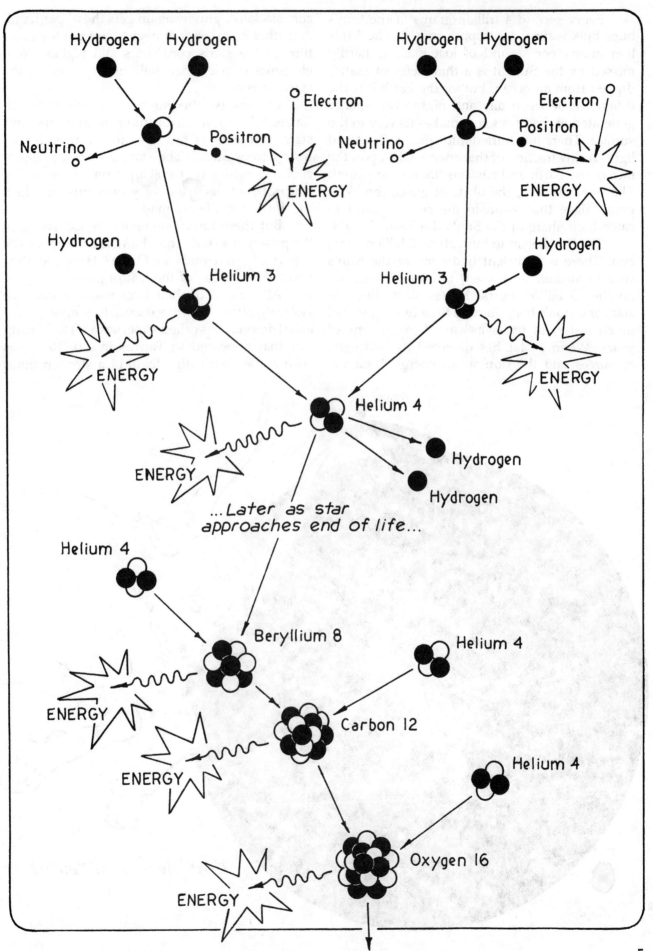

Hydrogen   Hydrogen

Electron

Neutrino

Positron

ENERGY

Hydrogen

Helium 3

ENERGY

Hydrogen   Hydrogen

Electron

Positron

Neutrino

ENERGY

Helium 3

Hydrogen

ENERGY

Helium 4

ENERGY

Hydrogen

Hydrogen

...Later as star approaches end of life...

Helium 4

Beryllium 8

ENERGY

Helium 4

Carbon 12

ENERGY

Helium 4

Oxygen 16

ENERGY

Every second 4 trillion grams of the Sun's huge bulk is turned into pure energy. The 4 trillion grams per second of lost mass is hardly missed by the Sun, it is a thimbleful of matter dipped from an ocean, but for the Earth it is the difference between day and night. The energy generated at the Sun's core makes its way to the surface. There it is hurled into space as heat and light. A tiny fraction of that energy falls upon the face of the Earth and sustains life on the planet. The tree, the fern, the blade of grass, the blue-green algae that swim in the sea, all bend to catch their share of the Sun's diminished mass.

The Sun began to burn about 5 billion years ago. There is sufficient hydrogen at the Sun's core to sustain its present rate of burning for another 5 billion years. Hotter stars, like the stars of Orion's belt, burn up their hydrogen fuel much faster, in times measured in millions of years. When a star has depleted its hydrogen resources and the outflow of energy from the core slackens, gravity again gets the upper hand. A further squeeze now begins and the temperature of the star's core soars still higher. Now elements heavier than helium are fused in the starry furnace.

In stars like the Sun, carbon will be the ultimate ash of nuclear burning. In more massive stars, elements as heavy as iron can be forged from lighter nuclei. The stars are great pressure cookers, kettles that boil up from a simmering soup of hydrogen the very elements of which the Earth and life are made.

But these heavy elements are locked up in the prison of a star's core. How did they get to be a part of gas clouds in Orion? How did they become the tissue of the leaf of grass?

At the end of their lives massive stars die violent deaths, blowing themselves apart in colossal detonations called supernovas. The blazing star that appeared in Taurus in A.D. 1054 was such a star-death, the death of a star ten times

Crab Nebula in Taurus

more massive than the Sun. The explosion in Taurus was visible to the naked eye only for a matter of months, but today, a thousand years later, when astronomers turn their telescopes to Taurus they see the ragged envelope of gas which is the expanding shroud of the dead star. The object is known as the Crab Nebula.

In the enormous energies generated by supernovas, elements even heavier than iron can be fused from lighter elements. All of this nuclear debris—the elements fused in the normal burning of the star and the heavier elements produced in the star's violent death—are flung into space. As the Crab Nebula continues to expand, its matter will be distributed through interstellar space, to become part of the gas and dust clouds of the Galaxy, perhaps to become part of some future star or planet.

As we look at Orion in the night sky we are looking at the birth of stars in dusty corners of the Galaxy. As we look at the Crab Nebula in Taurus we are looking at the death of stars and the enrichment of interstellar space with heavy elements. The planet Earth, with its wealth of carbon and silicon and oxygen and iron, would not exist if generations of massive short-lived stars had not lived and died within the arms of the Galaxy before the Solar System was born. These earlier generations of stars cooked up the stuff of which the Earth is made.

Often I have sat on a certain hill in the west of Ireland and looked out across the dark Atlantic to the Skellig Rock, and above the Rock to the stars of the Hunter and the Bull. The stars burn as brightly today as they did on those summer mornings long ago when the Crab supernova transformed the constellations. I have tried to imagine what the monks on Skellig Michael thought of the new star that flared so briefly on the Bull's horn. Nothing they thought or dreamed could have been more remarkable than the truth.

The "new star" that blazed out in the morning sky in A.D. 1054 was in fact an old star, a dying star, a star in the convulsive throes of self-destruction, dispersing into space the atoms of future stars and planets. The Earth is made of star-ash. We are ourselves made of star-ash. You and I and the leaf of grass are the journey-work of stars.

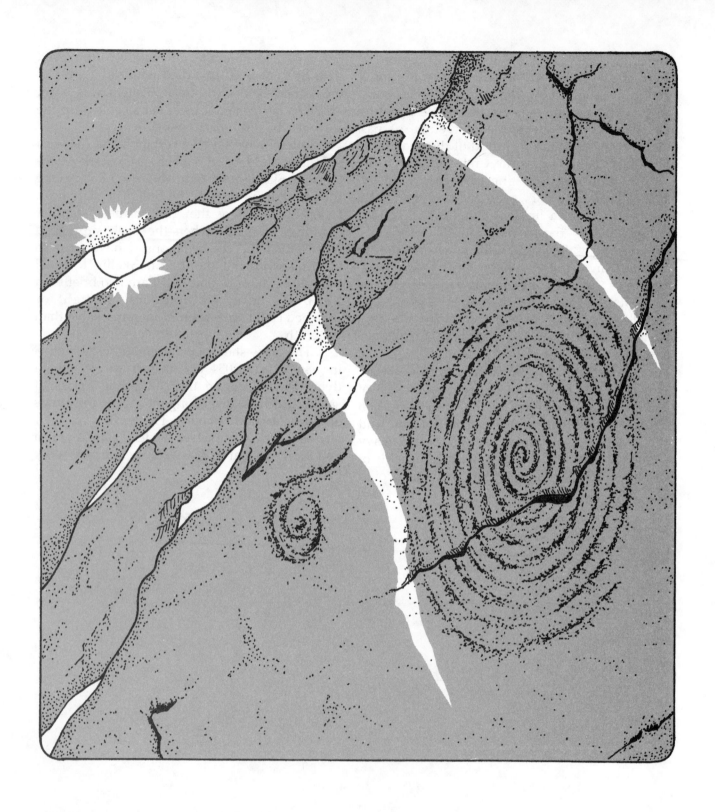

The Supernova Trigger

The catastrophic death of a massive nearby star may have triggered the formation of the Solar System.

The Anasazi were the ancestral people of the modern Pueblo Indians. They lived among windswept mesas, overhanging cliffs, and sun-baked canyons of the American Desert, near the place where Utah, Colorado, Arizona, and New Mexico share a common corner. Conditions for life in the Anasazi country were tentative. The lowlands were too dry for the growing of corn—the staple of the Anasazi diet—and the highlands were too cold. Drought was a recurring problem.

Yet the Anasazi thrived. Perhaps the best known relic of their civilization is the city built in the face of a cliff at Mesa Verde. But it was at Chaco Canyon in northern New Mexico that the Anasazi gift for organization and construction reached its zenith. Pueblo Bonito in Chaco Canyon was a monolithic structure of adobe and stone that contained 650 rooms. Other pueblos in the canyon added 2000 more rooms. Within and around these grand dwellings the Anasazi grew corn and ground it into flour, built irrigation ditches and dams, cut roads into the mesas, crafted splendid artifacts from turquoise, reeds and clay, and partook of elaborate ceremonials and rituals.

Life in Chaco Canyon depended desperately and delicately upon the Sun. The Anasazi attended to the Sun's passage with unceasing care. On these two pages I have illustrated an Anasazi observatory located on Fajada Butte in Chaco Canyon. Cracks between natural slabs of stone create "daggers" of light that fall on the shadowed face of the cliff. Spirals pecked into the cliff face by the Anasazi provide points of reference. As the "sun daggers" move across the spirals they record the Sun's journey across the sky and mark the cardinal points of the seasons.

The little sun-temple on Fajada Butte is just one of many solar observatories that have been identified among Anasazi ruins of the American southwest. The Anasazi were sunwatchers. They placed their hope, their confidence, in the Sun. That confidence was frustrated. By the 13th century a cooling climate brought cold and

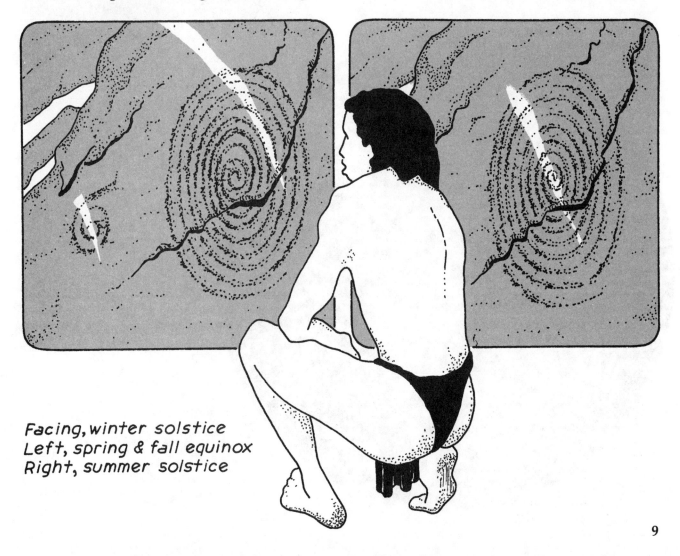

Facing, winter solstice
Left, spring & fall equinox
Right, summer solstice

drought to Chaco Canyon. The people moved away. The great pueblos fell into ruin. Today the canyon is a wasteland that belongs to the Sun alone.

Life is the gift of the Sun, to be given or taken away. If there are Earthlike bodies in the Universe that drift free of stars, we can be virtually certain they are devoid of life. It was the Sun's heat and light that struck the spark of life on planet Earth. It was the Sun's energy that fanned life's flame and sustained the great experiments of evolution. The story of life on Earth must begin with the story of the Sun.

The Sun had its origin nearly five billion years ago. It condensed with the Earth and other planets from an interstellar nebula of dust and gas. It now seems likely that a supernova may have presided at the birth of the Sun. It is a remarkable tale now to be told, and it will have a curious conclusion among the artifacts of the Anasazi.

A supernova is the explosive death of a massive star that scatters the star's substance into space. There are several hundred billion stars in the Milky Way Galaxy, and every century two or three blow themselves to bits. Most supernova events which occur in the Galaxy are obscured from our view by distance and by the dust and gas which clutter the arms of the star-studded Milky Way spiral. A few supernovas are near enough to be prominent temporary objects in Earth's sky. The Crab supernova that flared up in A.D. 1054 was one star-death that was noted on Earth. Several earlier events were recorded in Oriental, Arabic, and European chroni-

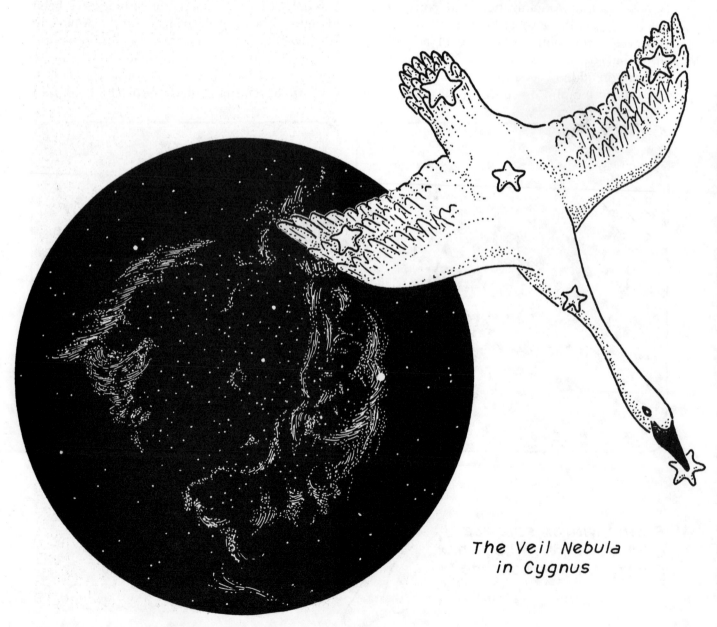

The Veil Nebula
in Cygnus

cles. The most recent supernovas to illuminate our night occurred in 1572 and 1604, and were studied by the Renaissance astronomers Tycho Brahe and Johannes Kepler. We are probably overdue for another supernova in our neighborhood of the Galaxy. Astronomers eagerly await the event.

The Crab Nebula in Taurus is the telescopic remnant of the A.D. 1054 supernova. Other bubbles and streamers of gas in the night sky have been identified as the remnants of supernovas of the historic and prehistoric past. At the left I have illustrated the hauntingly beautiful Veil Nebula in the constellation Cygnus the Swan. The Veil Nebula is about the same distance from Earth as the Orion Nebula and in the same spiral arm of the Milky Way as our Sun. It encircles a part of the sky six times bigger than the moon, but it is much too faint to be seen with the naked eye. Telescopic photographs reveal wisps and filaments of fluorescent gas a necklace of glowing matter.

The Veil Nebula began its life 40 thousand year ago when a massive exhausted star exploded, scattering its ashes into space. The event was certainly observed on Earth by our Cro-Magnon ancestors, crouching at the mouths of caves, marveling at the sudden transformation of their sky. The star that exploded was four times closer to us than the star which gave birth to the Crab. The "new Star" in Cygnus, the ancestor of the Veil, must surely have dominated Earth's night for many months.

At the instant of detonation, the matter ejected in the Cygnus supernova rushed outward at speeds of thousands of miles per second. In the fury of the explosion, the star blazed with dazzling intensity. An entire chemical stockroom of heavy elements was fused from lighter nuclei.

After a thousand years, the bubble of exploded stardust had swollen to the size of the Crab Nebula. By that time the expanding shock wave had swept past the stellar neighbors of the shattered star. As we look at the Veil Nebula today, it has expanded to become a gossamer sphere nearly 100 light-years in diameter. The bubble may now enclose several thousand stars.

As the shock wave rushed outward, it gathered up dust and gas from the space between the stars. This "dust-broom effect" is evident in photographs of the nebula. More background stars can be seen through the Cygnus

Supernova remnant in Canis Major

* New stars

loop than through the space around it. The loop is a little cosmic window, a volume of space between us and the background stars that has been swept clean of obscuring matter.

In their outrush, the shock waves of supernovas collect and concentrate interstellar dust and gas. If the shock wave of a supernova encounters a particularly dusty volume of space—the dark clouds in Orion, perhaps—the result can be dramatic. It is now widely believed that supernova shock waves can trigger the formation of new stars!

The drawing on the previous page shows a supernova relic in the constellation Canis Major, Orion's faithful dog. The present diameter of this ghostly arc of star-debris and its velocity of expansion indicate an age of 800,000 years, many times more ancient than the Veil. The exploding star that produced this wreath of gas appeared in Earth's sky at about the time humans were learning the use of fire. Today the supernova remnant is 200 light-years in diameter and pushing its way into an especially dusty corner of the Galaxy.

Most of the stars we see in the direction of the Canis Major remnant lie in the foreground or background of the expanding star-shroud. But there is a cluster of hot young stars strung like pearls along the edge of the expanding nebula. I have marked some of them with asterisks. The origin of these stars seems to be related to the supernova event.

As the shock wave of the Canis Major supernova expanded into space, it swept up matter in its path like dust before a sweeper's broom. At last the gas and dust along the shock front was sufficiently dense for gravity to pull the material together into clumps. Those clumps became the stars we now see burning on the edge of the nebula.

If this scenario is correct, the death of one massive exhausted star was the trigger for the birth of a cluster of new stars. In the process, the expanding supernova remnant endowed the new stars and their planets with heavy elements created by the exploded star during its lifetime of normal burning, and with still heavier elements fused in the violence of detonation.

Some of the hot new stars on the fringe of the expanding Canis Major nebula will themselves, after some millions of years, "go supernova." The entire process can repeat itself in a chain reaction of star-building. There is a "super-bubble" in Cygnus, a gigantic sphere of hot gas and dust filling almost the entire constellation. Some astronomers think the super-bubble was puffed up by a chain reaction of many supernova explosions, in just the way you might blow up a balloon with many short breaths. Other astronomers have gone so far as to suggest that the spiral form of our rotating galaxy may be partly due to chain-reaction sequences in the deaths and births of stars.

If supernovas can trigger the births of stars, it is reasonable to ask if a supernova explosion 4.6 billion years ago was responsible for the formation of our own star, the Sun. Some astronomers say yes. They cite evidence that literally fell from the sky. The evidence involves the isotopic composition of meteorites. To understand the evidence, I must take you on a short digression.

The nucleus of an atom is an assemblage of two kinds of subatomic particles—positively charged protons and uncharged neutrons. It is the number of protons in the nucleus of an atom that decides the chemical identity of an element. Hydrogen has one proton, helium two, lithium three, and so on. Isotopes are different forms of the same element, differing only in the number of neutrons in the nucleus of the atom. Oxygen 16, for example, has 8 protons and 8 neutrons in the nucleus. Oxygen 17 has 8 protons and 9 neutrons. Oxygen 18 has two extra neutrons. The extra neutrons do not effect the chemical nature of the element.

On Earth, the relative proportions of the different isotopes of an element never vary. For example, in any sample of oxygen 99.756 percent of atoms are oxygen 16, with 8 protons and 8 neutrons. A further .205 percent is oxygen 18 and the tiny remainder is oxygen 17. It doesn't matter if the oxygen is collected from the atmosphere above New York City or from the rocks of Australia, the mix of isotopes is constant. The same relative proportions of isotopes are found in rocks the astronauts brought back from the Moon. Of course, this is just what we would expect if the Earth and its moon formed from the same well-mixed store of materials.

The Solar System condensed by gravity 4.6 billion years ago from an interstellar cloud of dust and gas, the so-called presolar nebula. The cloud was mostly hydrogen and helium, but contained a smattering of heavier elements such

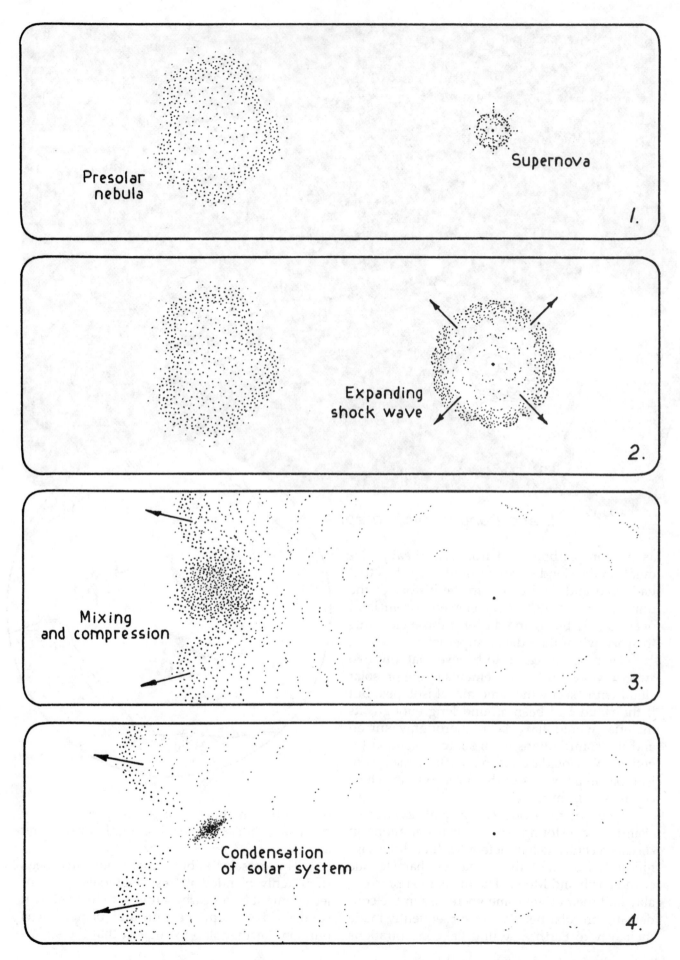

Chaco Canyon pictographs

as oxygen, carbon, and iron. The heavy elements in the cloud were contributed by stars that had lived and died earlier in the history of the universe. Stars fuse heavy elements from light ones as they burn, and disperse those elements to space when they die as supernovas.

There is no reason to believe that any two supernovas contributed elements to the presolar cloud with exactly the same mix of isotopes. But if the cloud had been around long enough, its material would have been thoroughly stirred and the contributions of all sources blended together. We would expect to find the same mix of isotopes in all bodies of the Solar System which condensed from the cloud.

Now for the evidence. Several researchers claim to have found isotopic concentrations in certain recently fallen meteorites that differ significantly from the mix of isotopes characteristic of the Earth and Moon. The meteorites presumably had their origin somewhere else in the condensing presolar nebula and subsequently made their way to Earth. The unusual concentrations of isotopes in the meteorites suggest that the presolar nebula was not thoroughly mixed after all.

The presolar nebula might not have been thoroughly blended if fresh materials were injected into it by a nearby supernova just before it condensed to form the Sun and planets. It is only one further step to suggest that the super-

14

nova which injected the fresh assortment of isotopes was also the trigger that caused the cloud to collapse. The hypothetical sequence of steps is diagramed on page 13.

Life on Earth is the child of the stars. We are made of the stuff of stars, of the ash of star-shine. The atoms in our bodies were forged in stars and blasted into space by supernovas millions or billions of years before the earth was born. And now, out of the sky fall chunks of stone that carry a startling message of the Earth's beginning—a supernova was midwife to the arrival of our planet on the cosmic scene.

The Anasazi people who watched the Sun with such interest from Fajada Butte must have witnessed the supernova of A.D. 1054, the super-nova of the Crab. Certainly the residents of Chaco Canyon lived in a more favorable latitude and had clearer skies than the monks on Skellig Rock in the North Atlantic. Unlike the monks, the Anasazi may have recorded the event.

The drawing above reconstructs the view from the rim of Chaco Canyon on the morning of the first appearance of the "new star," July 5, 1054. The tipped-over vee of stars at upper right is the head of Taurus the Bull. The supernova is the brilliant "star" that makes a little triangle with the crescent moon and the horn-star of the Bull.

The view in my drawing is toward the northeast, and already the dawn is beginning to brighten the sky. The conjunction of an eyelash-thin moon and a new star of dazzling brightness on that special morning a thousand years ago must have presented a spectacular aspect to anyone watching from the mesa above the canyon.

On an overhanging ledge of rock in Chaco Canyon there is a pictograph (rock drawing) that may record the supernova. I have sketched it here. The moon and the new star are evident. The hand print may be a signature of the artist.

There is no way to know if this pictograph in Chaco Canyon is exactly contemporary with the new star in Taurus. It is one of several moon-star rock drawings in the American West that date from approximately the time of the supernova.

The Chaco pictograph is at the opposite end of the canyon from the observatory on Fajada Butte. Together the two sites confirm an Anasazi awareness of the sky. The Anasazi knew, and knew truly, that the sky was the giver of life. We are the children of the Sun. We are the grand-children of the stars.

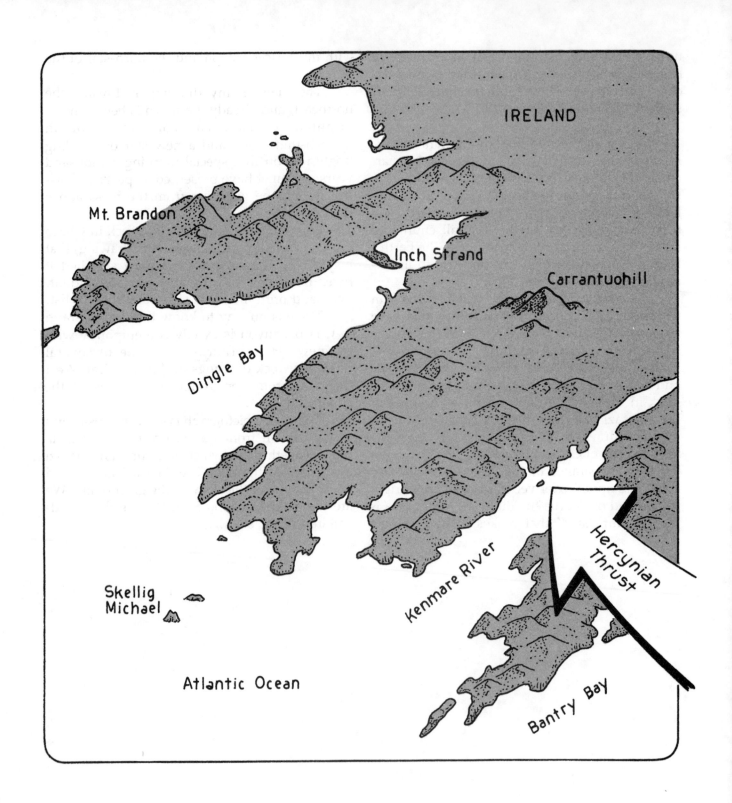

World Enough and Time

Two physical conditions are necessary for life to appear and prosper anywhere in the Universe. The first is a container with a suitable chemical and thermal environment. The second is lots of time.

The ingredients of life on Earth were collected by gravity. The hearth that held the tinder and received the spark of life was a small heavy-element planet near a yellow star. Chemistry was the steel and time the flint that struck the spark of life. For the spark to catch and the flame to grow required not biblical days, but hundreds of millions of years. The Solar System has been around for four and a half billion years. That's time enough for miracles.

One day last summer within a span of a few hours I vividly experienced the wealth of time allotted to life on Earth. In the morning I walked on Inch Strand, a promontory of yellow sand that nearly closes Dingle Bay in the west of Ireland. The tide was out and the sand bars were corrugated with ripples. In the afternoon I was on the summit of Carrantuohill, the highest peak in Ireland. On the shoulder of the mountain, three thousand feet above the sea, I stood beside a vertical slab of sandstone, marked with the very same corrugated ripples I had seen at the edge of Dingle Bay. The stone ripples on the summit of the mountain seemed as fresh as if they had been created that morning by the ebb and flow of the sea.

My geology map of Ireland tells me that the sandstones on Carrantuohill date from the Devonian period, 400 million years before the present. The ripple marks on the vertical slab of stone were convincing evidence that the rocks at the mountain's summit had once been sand swept by the flow of the tide. How then, was that ancient beach turned into stone and thrust 3000 feet into the air?

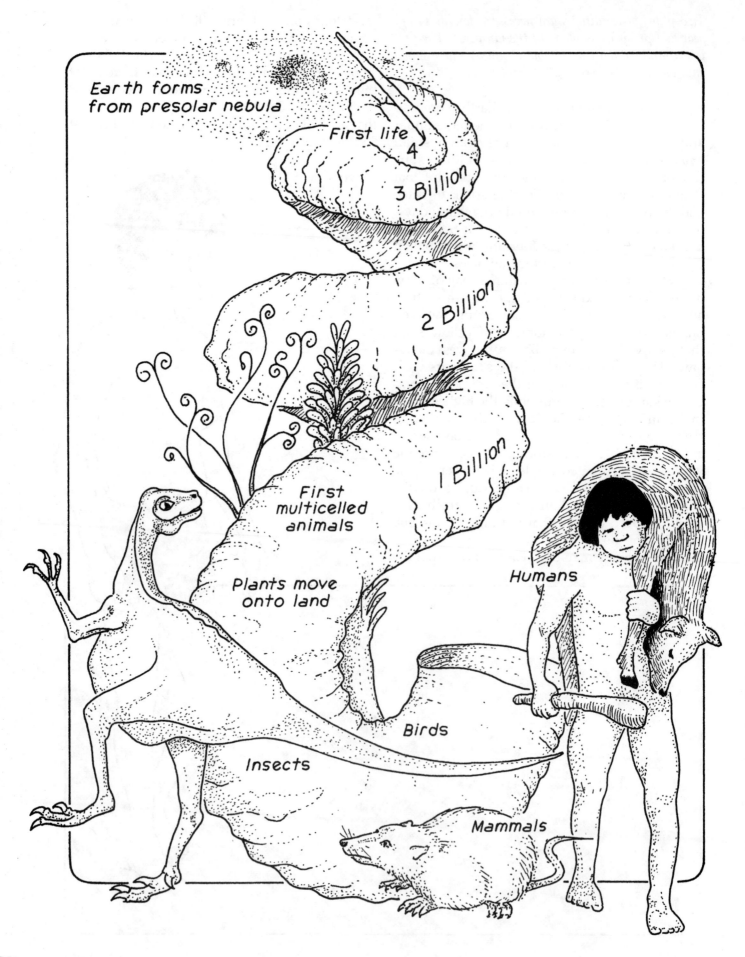

Earth forms
from presolar nebula

First life
4
3 Billion

2 Billion

1 Billion

First
multicelled
animals

Plants move
onto land

Humans

Birds

Insects

Mammals

According to geologists, the rocks at the summit of Carrantuohill are part of a series known as Old Red Sandstone. The formation consists of sandstones mixed with some mudstones and pebblestones laid down in Devonian times in inland lakes or shallow inlets of the sea.

Within the formation can be found the fossils of early fish, crayfishlike anthropods, and early land plants. The fossils in the Old Red Sandstone were among the first evidences of ancient life to be discussed publicly by geologists of the early 19th century. The fossils helped convince a skeptical public that life had been present on the planet long before humans trod the Earth.

The sands and muds deposited in Devonian lakes and tidal basins were buried deeply by successive layers of sediment. Ultimately the sediments were turned to stone by pressure and chemical cementation. Then, between 300 and 350 million years ago, the rocks of northern Europe were crumpled up into high ranges by a squeezing pressure from the south. The squeeze is known as the Hercynian Orogeny (an orogeny is an episode of mountain building). Carrantuohill is the eroded stump of one of those great Hercynian peaks. Hundreds of millions of years of erosion have pared down the summit and exposed at last the ancient rippled beach that had lain so long hidden in the heart of a mountain.

So come stand with me at the base of the corrugated stone slab on the summit of Carrantuohill high above the shining strand at Inch, and reflect upon the hundreds of millions of years which separate the living beach from the fossil beach. Through all that time life changed itself in marvelous ways. Plants moved from shallow margins of the sea onto the land. Great forests colonized continents. Fish crept ashore and became amphibians. Amphibians became reptiles. Dinosaurs pushed forward to dominate the planet and then fell precipitously from prominence. Mammals moved to occupy the niche vacated by the vanished dinosaurs. Grasses spread a carpet of green across the land and herds of hoofed beasts appeared to roam the grasslands. Primates stood up on their hind legs and became human. Four hundred million years of astonishing inventiveness—the seed, the feather, the flower, fur, the grasping hand. And all that time Carrantuohill was going up inch by inch and coming down grain by grain.

In my drawing, I have represented the long sweep of geologic time by the coiled shell of a fossil mollusc of late Cretaceous times (70 million years ago). This particular mollusc is perfectly suited for its role as a time line, with one multichambered coil to represent each of the billion years of Earth's history.

A billion years! What is a billion years to a creature that measures its life in hours and minutes? Make each of those billion years a grain of sand and you've got a beach.

It was James Hutton, gentleman farmer of Scotland, who taught us how to look backward along the giddy spiral of time and see more than the several thousand years allotted to the Earth by biblical history. Hutton's *Theory of the Earth*, read before the Royal Society of Edinburgh in 1785, conferred upon the planet sufficient time for nature to raise mountains up and tear them down without the agency of specific and recurring divine intervention. Hutton's treatise was subtitled "An Investigation of the Laws Observable in the Composition, Dissolution, and Restoration of Land Upon the Globe." The subtitle is a mouthful, but it simply means that the history of the Earth is written in the record of the rocks—for those who will read it.

As he studied the rocks of his native Scotland—among them the Old Red Sandstone I found on Carrantuohill—Hutton saw in his mind's eye the beaches that had been buried and turned to stone, and the mountains that rose and fell like waves on the sea. He saw, he said, "no vestige of a beginning, no prospect of an end" for the forces that had shaped the face of the Earth. Hutton's gift to science was the gift of time, geologic time, rock time, time in which all of human history was but the tick of a clock.

In a stream bank near Jedburgh in southern Scotland, Hutton found an assemblage of sedimentary rocks that spoke eloquently of the Earth's antiquity. At the bottom of the bank Hutton observed vertical strata of sedimentary rock that had been tipped up from their original horizontal position. On the eroded tops of these vertical strata lay horizontal beds of Old Red Sandstone (see drawing on page 22).

The strata tell their own story. First, the lower strata were deposited in an ancient sedimentary basin. Then, a contraction in the Earth's crust folded or tilted the strata into a vertical posture, perhaps heaving them into high mountains above the level of the sea. Erosion

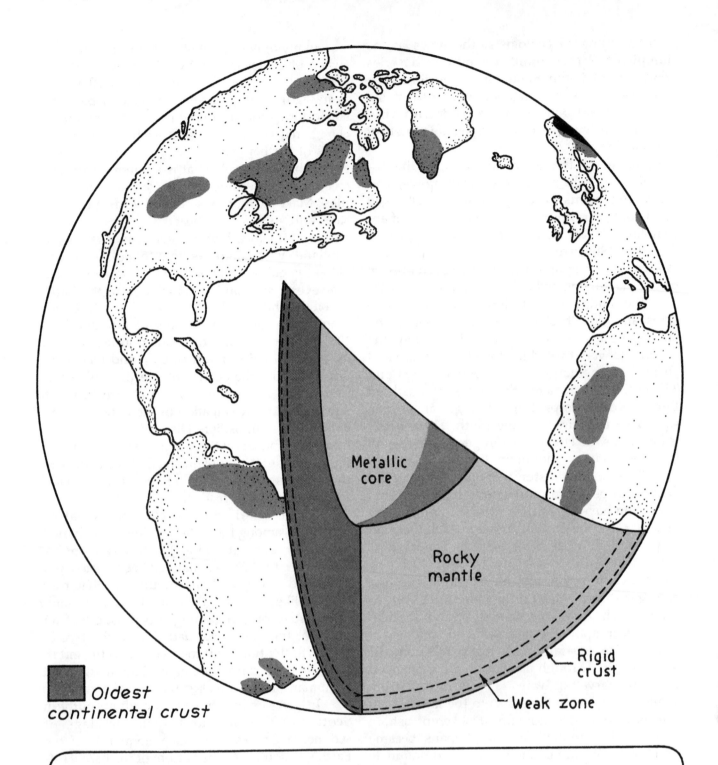

Oldest continental crust

EARTH DATA

Radius of the Earth: 3958 miles/6368 kilometers

Inner core is solid iron and nickel.
Radius of inner core: 756 miles/1216 kilometers

Outer core is molten iron and nickel.
Thickness of outer core: 1376 miles/2200 kilometers

Mantle is composed of solid silicate rocks
(close to melting point near top).
Thickness of mantle: 1800 miles/2900 kilometers

Continental crust is composed of granitic rocks.
Thickness of continental crust: 10–40 miles/16–64 kilometers

Oceanic crust is composed of basaltic rocks.
Thickness of oceanic crust: 7 miles/11 kilometers

Continental crust is 40% of the surface.
Oceanic crust is 60% of the surface.

went to work, tearing the mountains down, at last planing off the tops of the vertical strata. The crust sagged, the platform of vertical strata dropped, the sea intruded upon the land, and new layers of sandstone were deposited across the eroded stumps of the ancient rocks. Again the Earth's crust lifted, this time without significant tilting, pushing the rocks above the level of the sea where the agents of erosion could begin to cut them away. This was the story of the rocks observed by Hutton in the stream cut near Jedburgh.

Geologists would now call the interface between the vertical and horizontal sedimentary strata an unconformity. It represents a clear break in the pattern of deposition of sediments, episodes of geological violence that have been removed by erosion from the record of the rocks, like pages torn from a book.

Water rushing through the countryside had opened the book of the Earth's crust, and Hutton was there to read it. He carried with him a set of "spectacles" that enabled him to read a story which others had failed to read.

Hutton's "spectacles" were the idea of the yawning, inhuman gulf of geologic time. In the strata exposed along the stream bank, conjoined in so unlikely a fashion, Hutton saw the long history of the Earth. He deduced from the rocks laid bare before him the deposition of sediments, their consolidation into stone, their uplift and deformation, erosion, subsidence and more deposition, more uplift and more erosion, mountains rising and falling, heaving up and tumbling down, page on page, chapter on chapter of a story which once read changed forever the way we regard the planet Earth.

Far back in the "annals of the former world" (Hutton's phrase), on the first page, so to speak, we read of the formation of the Earth from the dust of space. Mystery and controversy surround the Earth's beginnings, but virtually all astronomers and geologists agree on the central plot. The Earth, the Moon, the Sun, and the other planets condensed together from a cloud of interstellar dust and gas, the presolar nebula. The cloud was mostly hydrogen and helium, but contained a smattering of heavier elements like carbon, silicon, oxygen, and iron. A nearby supernova may have triggered the collapse, but it was gravity that pulled the cloud together. Gravity is the elastic which pervades the Universe and—unopposed—causes everything to fall into clumps.

The nebula which became the Solar System had some small degree of rotational motion. As the nebula compacted under the influence of gravity it began to rotate more rapidly. Physicists call this effect the "conservation of angular momentum," but we can call it the "ice-skater effect." Just as an ice skater spins faster as he pulls his arms closer to his body, so did the presolar nebula spin faster as gravity packed its mass more tightly near the axis of rotation.

As the cloud spun faster it tended to flatten out. Physicists refer to "centrifugal force." We can think of a pizza chef spinning a clump of dough and call it the "pizza effect." The flattened disk of hydrogen, helium, and supernova debris became the Solar System.

Most of the matter in the flattening cloud was pulled to the center of the disk and became the Sun. Gravity continued to squeeze the embryonic Sun, and the temperature soared. When the temperature at the core of the central sphere reached 8 million degrees fusion began, and the Sun turned on in a blaze of glory.

Meanwhile, in the spinning disk, atoms and molecules condensed to form dust-sized grains. The composition of the grains depended on the temperature in that part of the disk where they formed. In the hot inner part of the disk, close to the new Sun, only rocky and metallic elements were massy enough for gravity to clump them together, against the tendency of thermal motion to keep them dispersed. The metallic and rocky grains became the stuff of the inner planets—Mercury, Venus, Earth, and Mars.

Further out in the whirling disk, where temperatures were lower and thermal agitation less vigorous, light elements and volatile icy compounds condensed and ultimately contributed their mass to the huge, low-density outer planets.

As time passed, the metallic and rocky grains in the inner part of the solar disk clumped together into asteroid-sized bodies—perhaps a mile in diameter—and these in turn collected by collision and the pull of gravity to form the early planets. At some point in this process, the Sun turned on with a sudden violence that swept the inner Solar System clear of the lighter leftovers. Whatever gases clung to the inner planets were blown away.

All of this took place some 4.6 billion years ago.

There is considerable controversy over

what happened next. Perhaps the Earth had a cold beginning, growing like a big dirty snowball from clumps of whatever materials had condensed at the Earth's distance from the Sun. Or perhaps the Earth was hot at its birth and there was some layering of materials even as the planet formed. In any case, the planet soon heated up, melted, and sorted itself out by density, with the heaviest materials sinking toward the center of the Earth and the lightest materials rising toward the surface (more on this in the next chapter). Within half a billion years of its formation, the Earth was physically similar to the Earth today. It had a metallic core, a rocky mantle, and a crust that had cooled sufficiently to provide a solid platform for the great experiment of life. The plant also had an atmosphere and an ocean.

A map of the surface of the primitive Earth would be very different from a map of the present globe. The rigid crust of the Earth is thin and fragile—as thin, relatively speaking, as the shell of an egg. The "eggshell" has been crunched, stretched, broken, and rearranged by geological violence. Crustal rocks have been recycled through the Earth's hot interior. The present continents are products of geological violence that has continued throughout the Earth's long history. The oldest rocks of the present continents (those more than 2½ billion years old) are shown on the drawing on page 20, but these rocks have been shuffled and reshuffled during the intervening eons and almost certainly give little hint of what the Earth's continental crust looked like at the time of its formation.

In such a way was the Earth prepared for the spark of life. In a sense, the outcome of the story was inevitable. Every atom in the universe is imbued with that attractive quality called gravity. Every atom has the chemical propensity to link up with other atoms in ways which lower the energy state of the combination. Even in its tiniest grain, the Universe is quickened with a drive toward consolidation and complexity. All that was required for the creation of a star, an Earth, and ultimately life, was time—Hutton's

Unconformity at
Siccar Point, Scotland

Scotland

England

time, the time of the rocks on Carrantuohill, time without vestige of a beginning or prospect of an end.

From his study of the rocks of Berwickshire in Scotland, James Hutton reasoned that somewhere along the coast of that county he would find an unconformity of strata similar to the one he had found in the stream cut near Jedburgh. With his friends James Hall and John Playfair he took a boat and moved along the coastal cliffs where the Lammermuir Hills of Berwickshire reach the sea. At a place called Siccar Point they found the formation Hutton was searching for, the formation illustrated here, Old Red Sandstone strata lying atop even more ancient "schistus" that had been tipped almost vertical and eroded flat.

According to Hutton, eons of time and long invisible episodes of uplift and erosion were required for the creation of this particular configuration of stone. The Earth was ancient, he declared, beyond anyone's knowing. John Playfair recorded Hutton's vision of deep, geologic time. He wrote: "On us who saw these phenomena for the first time, the impression will not easily be forgotten. . . . We felt ourselves necessarily carried back to the time when the schistus on which we stood was yet at the bottom of the sea, and when the sandstone before us was only beginning to be deposited, in the shape of sand and mud, from the waters of a superincumbent ocean. . . . The mind seemed to grow giddy by looking so far into the abyss of time."

A Warm
Little Pond

The chemicals of life collected on the surface of the Earth early in its history. They came from the Earth's interior or as passengers on meteorites.

Life is almost as ancient as the Earth itself. Certain fossil structures 3½ billion years old suggest that even then the Earth's surface was gauzed with living substance. Certainly, rocks 2 billion years old contain evidence for life that is unambiguous and pervasive. Those primitive life-forms were simple, to be sure, microscopic even, but already they were powerfully at work modifying the crust and atmosphere of the young planet.

Precisely how, when, and where life made its debut on Earth may never be known. But most contemporary scientists agree that the first living cells arose from spontaneous chemical arrangements of nonliving matter.

Even the simplest living thing—a single-celled bacterium, for example—seems marvelously advanced over the most complex nonliving structures. If ever there was a "missing link," it is the giant step between the shelf of dead chemicals and the living organism. That the animate should arise from the inanimate must always seem a bit of a miracle.

Why did the miracle happen only once, 3½ billion years ago, and not again since? In Darwin's day it was protested that if life came from nonlife at some time in the past, then we should see the same thing happening today. New creatures should spring from every fetid pool. Every rain barrel should clamor with exotic organisms. And, of course, such things do not happen. In our experience, life—always, invariably, reliably—comes only from life.

Darwin had a wise reply: If a simple form of life were to appear spontaneously today, it would immediately be devoured or absorbed by living creatures. In our present environment, life is pervasive and voracious. A fresh starter would never have a chance.

What is required for the spontaneous genesis of life, said Darwin, is a "warm little pond," rich with the right chemicals and free of living predators. Darwin's "warm little pond" was the planet Earth about 3½ billion years ago.

More precisely, the "warm little pond" was Earth's thin skin of water and air. If the Earth were the size of this page, the oceans and atmosphere would occupy a layer no thicker than this piece of paper. But without that gossamer film of fluids the Earth would be as lifeless as the Moon.

When I think about an Earth without its oceans and atmosphere, two images come to mind. Both images involve the Moon.

I have often watched an occultation of a star or planet by the Moon. An occultation occurs when the Moon, in its monthly circuit of the Earth, passes in front of another celestial object. On the facing page I have sketched an occultation of the planet Venus that occurred on the morning after Christmas, 1978—as best I can reconstruct if from notes in my journal. The image is inverted as it appeared in my telescope. As the Moon passed over the tiny crescent of the

planet there was no blurring, no refraction, no distortion of the planet's disk. The Moon's limb sliced across the face of the planet like the edge of a razor. No further proof should be needed that the Moon has no atmosphere, no fuzzy skin of air to blur the extinction of the star.

The second image seems to contradict the first. It is the famous photograph taken by Apollo astronauts of the American flag waving in a lunar breeze. A breeze? On an airless Moon? Of course, there is no breeze at all. The waving flag is an illusion. NASA scientists had the foresight to provide the astronauts with a crinkly flag that could be held extended by a slender rod. A true flag would be a useless accessory on the Moon, forever dangling at its staff. The Moon has no buoyant skin of air. Gravity rules. Objects lie where they fall. Nothing wafts. Nothing floats. If the molecules of life existed on the Moon, they would remain forever dispersed, fixed in place for the eons like grains of lunar dust.

The Earth may once have been as airless as the Moon. It might be dry and airless yet except for one extraordinary event: Early in its history the Earth melted!

The Earth is partly molten still. The outer part of the core is molten. A layer of the mantle just below the crust is close to the melting point. Punch a hole in the rigid crust of the Earth and red-hot liquid rock would ooze to the surface.

The Earth was warm at its formation. The squeeze of gravity, radioactivity, and the bombardment of asteroids heated it further. What happened next has been hotly debated by geologists, but a possible scenario follows. A substantial part of the Earth's bulk is iron, and of the major materials that make up the Earth iron is the heaviest and has the lowest melting point. When the temperature of the warming Earth reached the melting point of iron, the iron liquefied and fell toward the center, displacing lighter rocky materials. This "big burp," this huge turnover of the Earth's substance— sometimes called the "iron catastrophe"—may have been the most momentous event in the

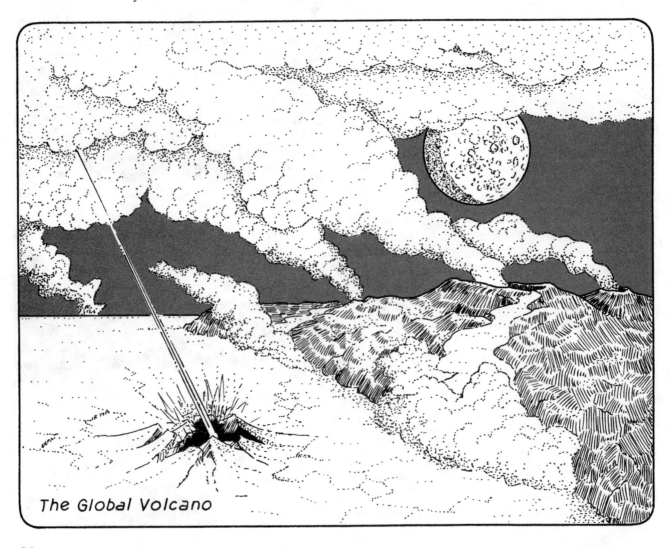

The Global Volcano

planet's history. The falling iron, like a pile driver, released gravitational energy in the form of still more heat. The planet melted. Perhaps it melted completely. Perhaps it became a big red liquid bubble of molten iron and rock.

After the "iron catastrophe" the surface of the young planet may have looked like the churning fire pit of Hawaii's Kilauea volcano, filled with lava lakes that crust over and quickly melt again. Eventually, as the planet cooled, a permanent crust began to accumulate from light rocky minerals that made their way to the surface of the partly fluid planet. Broad lava seas lay between cooler uplands. Those uplands became the cores of future continents.

Inside the hot plant a separation of materials was going on according to density. The heavy metallic compounds sank toward the core. Lighter materials tended to rise. The lightest of all, the volatile compounds, the water and gases that were chemically bound or otherwise trapped in the body of the planet, escaped to the surface. They bubbled from the surface of the

molten Earth even as they do today from volcanoes, geysers, and hot springs. The water, of course, was in the form of steam. The surface of the planet was still too hot for water to exist as a liquid.

In my sketch of the global volcano, I have shown the Moon looming in the background, abnormally large. Four billion years ago the Moon was closer to the Earth than it is today. Like the Earth, the Moon was volcanically active, glowing eerily in the night sky with its own red light. Steam and gases emerged from lunar volcanoes, as they did from the volcanoes of the Earth. But, like little Mercury, the Moon lacked sufficient gravity to bind gases to its surface and so never acquired an atmosphere.

As the Moon's crust formed, it was heavily bombarded by meteorites and asteroids, some as large as the state of Rhode Island. This ubiquitous early cratering is still much in evidence. Anyone who looks at the pockmarked face of the Moon through a telescope must wonder how the Earth escaped the terrible cosmic pummeling

The Ubiquitous Bombardment

that so battered our satellite. The answer: the Earth did *not* escape.

The Earth and Moon formed from the coalescence of smaller bodies that had condensed from the presolar nebula. Leftovers of that chunky cloud continued to rain down on the young Earth and Moon for a long time after their formation. This rain of meteorites slackened rather abruptly about 3.9 billion years ago, with one last great shower of stone—a "terminal cataclysm"—that made swiss cheese of the surfaces of the Earth and Moon. Other bodies in the Solar System also received a saturation bombardment. The scars of that fierce cratering are still evident on the smaller bodies of the Solar System. On the geologically active Earth the record of bombardment has been erased.

Small bodies cool more rapidly than large ones. The Moon's crust cooled quickly and was soon cold and rigid to a depth of hundreds of miles. The Earth's crust, on the other hand, is even today only tens of miles thick, and rests on a hot interior that is plastic and turbulent. Over the billions of years the Earth's eggshell-thin crust has been shifted, crumpled, squashed, rearranged, and made anew by forces stirred by the planet's internal heat. To this internal violence add the eroding effect of the Earth's atmosphere and oceans. The crust has been scratched, clawed, disassembled, and redistributed by water and weather. As a result of all this activity, no scars remain on Earth of the early bombardment. The Moon's cratered surface is a kind of museum that preserves an image of the violence that was once Earth's.

But some reminder of the early shower of stone may be with us yet. The meteorites and asteroids that fell onto the new planet carried trapped liquids and gases. This would be particularly true if the projectiles had their origin farther out in the cooler regions of the Solar System, where gravity was able to get a grip on volatile substances. In this way, at least part of the Earth's oceans and atmosphere came from outside, as passengers on meteorites. The rain of

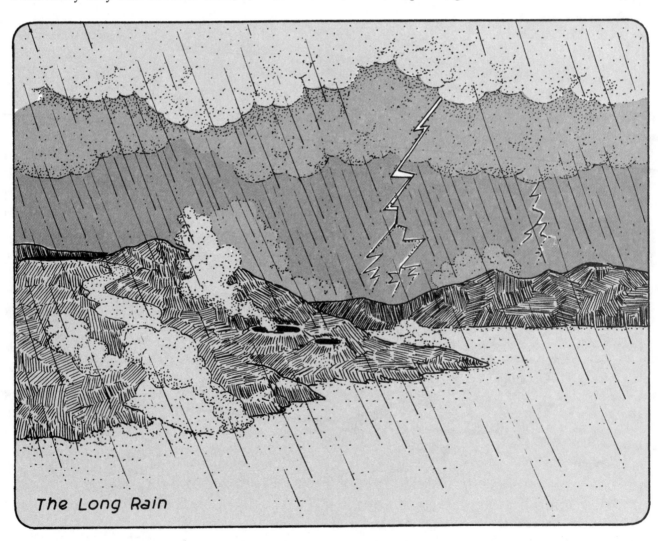

The Long Rain

28

meteorites may also have carried to the Earth's surface some of the molecular constituents of life.

The Earth's early atmosphere had a composition very different from the atmosphere today, a circumstance that will have important consequences for the story of life. The main gases contributed to the early atmosphere by volcanic activity were carbon dioxide, nitrogen, water vapor, carbon monoxide, and hydrogen. The hydrogen was too light to be gravitationally bound to the planet and escaped to space. Oxygen was present in combination with other elements, but there was no free oxygen in the primitive atmosphere. The absence of oxygen was the most significant difference between the Earth's early atmosphere and the present atmosphere.

As the Earth's crust cooled, the water vapor in the atmosphere began to condense into droplets and fall as rain. At first, when the rain fell onto the hot crust it boiled off as steam, like drops of water flung onto a hot griddle. The steam rose into the atmosphere, condensed, and fell again. And so it rained. It rained everywhere on the hot cloud-darkened planet. For thousands of millions of years there was not a single sunny day, not one starry night. Lightning crackled continuously. The downpour washed from the sky volcanic ash and the dust blasted into the atmosphere by meteorite bombardment. Eventually the surface cooled to the point where the water was able to remain liquid. It sizzled and steamed on the flanks of volcanoes. It cascaded into the lowland basins and the bowls of impact craters. At last, long last, the skies began to clear and the sun glistened on a sparkling sea. Among the most ancient rocks yet discovered on this planet are 3.8 billion year old metamorphosed marine sediments. By 3.8 billion years ago (at least) the Earth had acquired an ocean!

An exact timetable of the events I have described here is not known. Each of my three drawings emphasizes one phenomenon—volcanic activity, meteorite bombardment, a global rain—but they almost certainly overlapped. When the skies cleared, the volcanoes quieted, and the cosmic bombardment slackened, the Earth was encased with a film of air and water. It was still a bleak and barren place, still warm from its birth and bathed in ultraviolet radiation from the Sun. But as the violence from above and below subsided, it was undeniably home.

And so was Darwin's "warm little pond" prepared for the coming of life.

Some scientists have suggested that life arrived on Earth from elsewhere, fully formed. It might have arrived by accident, they claim, as a passenger on a meteorite or comet. Or the Earth might have been "seeded" by an extraterrestrial civilization which realized a pond had been suitably prepared for a grand experiment.

But it hardly seems necessary to invoke an extraterrestrial origin for life. All of the elements required for life were present in the early oceans and atmosphere. Carbon, nitrogen, oxygen, hydrogen, phosphorous, sulfur, were available in one form or another for transformation by energy. Carbon monoxide and hydrogen, for example, could react to form methane. Nitrogen and hydrogen could form ammonia. More complex organic compounds would certainly have followed (more on this in the next chapter). The thrust was toward complexity. The tendency of matter to organize itself into complex structures, given the appropriate environment and abundant energy, is part of the miracle of matter itself. The stuff of this world is like a magical tinker-toy set that self-assembles.

There was plenty of energy available on the new planet to transform the chemicals of life into living organisms. The Sun's high-energy ultraviolet radiation was not yet screened by an ozone layer in the upper atmosphere (ozone is a form of oxygen). The atmosphere crackled with electrical discharges. Radioactivity in the rocks was more pervasive than today. Volcanoes and meteorites continued to supply energy on a lessened scale. The pond was stocked with the right ingredients. The pond was "warm" indeed.

Nothing, it seems, could hold life back. In our experience, there are two things that resist the tendency of matter to form complex organic molecules—decay and oxidation. Decay is the work of living organisms—the worm, the fungus, the microorganism. Oxidation is combination with oxygen, as when wood burns and iron rusts. But living organisms and free oxygen were precisely the two things that were not present in the "warm little pond" where life had its genesis.

Before life could begin, complex molecules like proteins had to appear spontaneously. When living organisms build proteins they are invariably aided by a special type of organic molecule called an enzyme. Enzymes are themselves proteins that play the role of catalysts,

speeding up reactions that unassisted would take inordinate amounts of time. Clearly there were no enzymes in the "warm little pond" to speed up the synthesis of the first proteins. But no matter, life was in no rush. There was plenty of time. The oceans and atmosphere were in place, charged with energy, poised for life, 3.8 billion years ago.

Perhaps no one has experienced more profoundly the special character of the Earth as a suitable environment for life than the few creatures who have left its surface. The astronauts who went to the Moon took a bit of the "warm little pond" with them, backpacks full of the fluids and gases which are the medium of life on Earth. Standing on the cold, dry, airless lunar plains they looked out into the black void of space and saw the planet Earth hanging from the sun by a gravitational thread, glistening in its delicate sheath of sea and cloud.

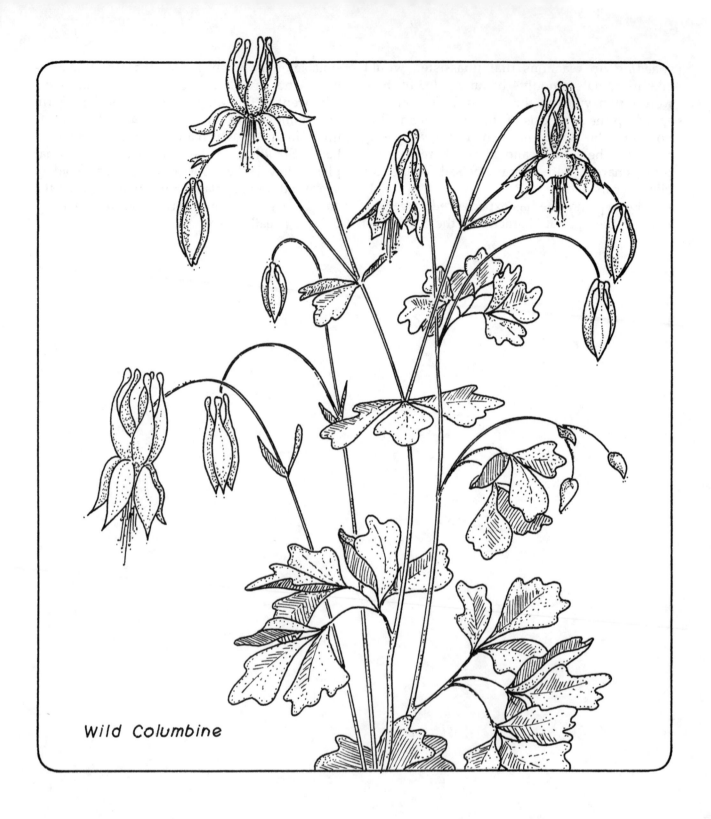

Wild Columbine

Fire Burn, Cauldron Bubble

The first living cells evolved from inanimate matter. Energy, water, and time were crucial to the process.

For almost twenty years I have ranged the woods and fields near my home in eastern Massachusetts. Most of the plants and animals have become familiar friends. By paying close attention to the weather, I can predict almost to a day when the first red-wing blackbird will reappear along the brook, or when the first cinnamon fern will unfurl its fiddleheads near the pond, or when the first russula mushroom will push its rosy cap up through the leaf litter on the forest floor. There is a pleasure in the familiar, in the recurring patterns of the seasons.

But there is also a special pleasure in the unexpected. Consider the lovely bell-flowered plant I have illustrated here, the wild columbine. Not a particularly rare flower I am told, but around here rare enough that in all my years of walking with my eyes open, I have seen only one plant. I found it growing on a mossy outcrop in the deep woods. Such an exquisite plant to find without precedent! It came, apparently, from nowhere, and left without progeny. A miracle, it seems, of creation from nothing.

No, not a miracle. Somehow a columbine seed had made its way to my mossy outcrop. That tiny package contained the blueprint for the finished plant. The orange five-part nodding blossoms. The trailing spurs. The golden stamens, showering like fireworks from the bells. They were all there, marvelously encapsulated in a tiny seed that somehow made its way to my outcrop like a cosmic space traveller.

The origin of my columbine was a mystery which is enfolded within the greater mystery which is the origin of life itself. That greater mystery has been poked and prodded by science, it has yielded up secret after secret, and yet the mystery remains as deep as ever.

I can be certain that my wild columbine

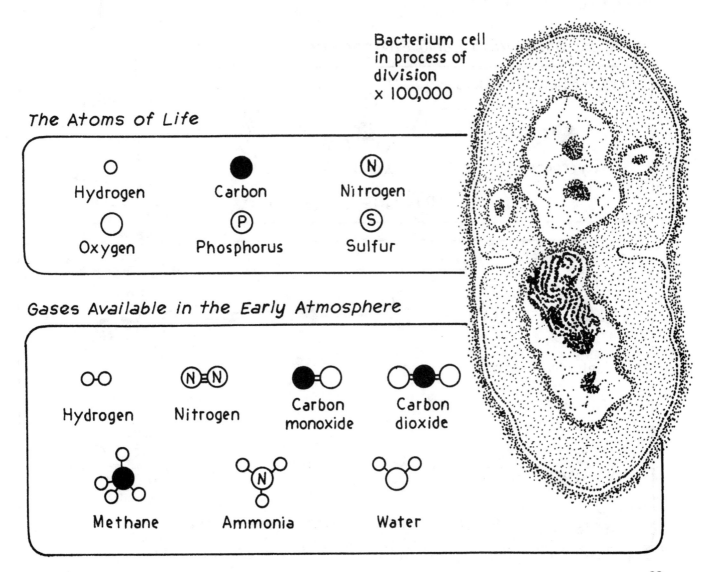

Bacterium cell in process of division x 100,000

The Atoms of Life

Hydrogen Carbon Nitrogen
Oxygen Phosphorus Sulfur

Gases Available in the Early Atmosphere

Hydrogen Nitrogen Carbon monoxide Carbon dioxide

Methane Ammonia Water

grew from a seed, and that the origin of the seed was another columbine plant in someone else's woods. But how did the first living thing come to be, the creature without a parent?

The first living creatures were single-celled organisms, perhaps similar to the simplest bacteria that exist today. Those primitive cells had the ability to survive over the short run by extracting energy from their environment, and to survive the long run by reproduction of their species. They were the first tentative steps on life's precarious journey.

The bacterium and the blue whale, the camel and columbine, all share the same chemistry under the skin. The molecular building blocks of life are surprisingly limited: carbohydrates, fats, proteins, and nucleic acids. With the exception of the last, the sort of things you find listed on the side of your breakfast food box. The original building blocks for life were fabricated from materials available in the early oceans and atmosphere.

On this page I have illustrated a version of an experiment first performed by Harold Urey and Stanley Miller in the 1950s. Ammonia and methane were introduced with water into a flask and subjected to a continuous electrical discharge. When the spark was turned off, the walls of the flask were coated with a scum of organic compounds. In particular, the amino

Molecule Building in Early Environment

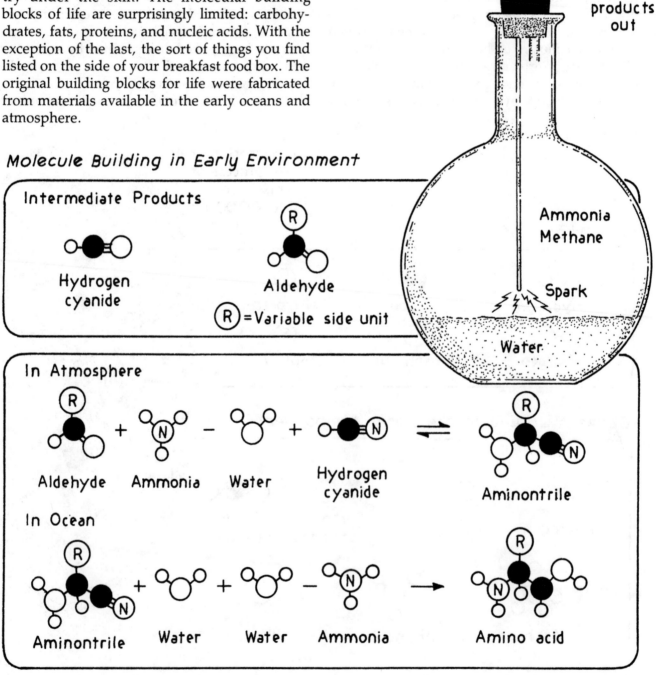

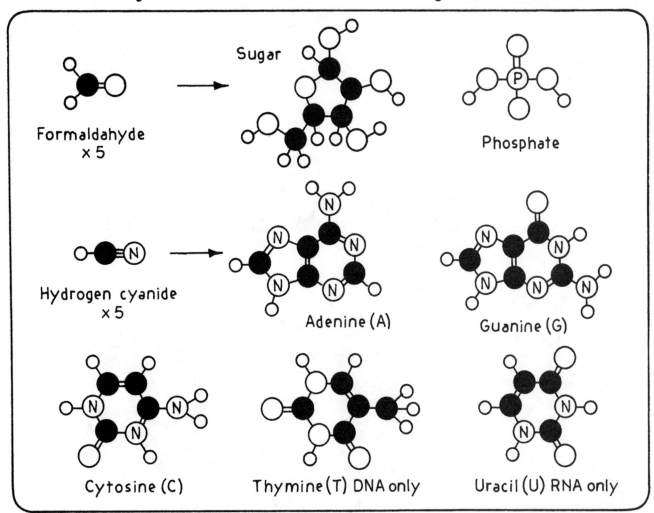

Formaldahyde x 5

Sugar

Phosphate

Hydrogen cyanide x 5

Adenine (A)

Guanine (G)

Cytosine (C)

Thymine (T) DNA only

Uracil (U) RNA only

acids, which are the subunits of the proteins, were found in the flask. Variations on the Urey-Miller experiment have been performed many times by other researchers. A wide variety of organic compounds have been fabricated.

The Urey-Miller experiment was an attempt to recreate the conditions that existed on the early Earth. The same chemical reactions that occurred in their flask can be presumed to have happened in the early environment.

Building organic molecules out of simple gases and water requires energy. In the Urey-Miller experiment, the energy source was an electric spark. On the early Earth the main sources of energy were ultraviolet light from the Sun and the crackle of lightning in the atmosphere.

Light from the Sun spans a spectrum of wavelengths, from the infrared, through the visible rainbow, to the ultraviolet. Only ultraviolet light carries sufficient energy to make or break the chemical bonds that hold molecules together. For example, it is ultraviolet sunlight that shakes up the molecules in your skin and gives you a tan. An ozone layer in the Earth's present atmosphere absorbs most of the Sun's ultraviolet radiation. And a good thing, too, or that lovely tan might give way to devastating molecular damage.

No ozone layer shielded the surface of the early Earth. Ozone is a form of oxygen. There was no free oxygen in the Earth's primitive atmosphere, and consequently no ozone. Ultraviolet radiation bathed the surface of the Earth and stimulated the formation of organic molecules—sugars, phosphates, amino acids, and organic bases. In a sense, the entire planet was a Urey-Miller flask, charged with energy and primed for the creation of simple organic compounds. But ultraviolet radiation can destroy complex organic molecules, such as proteins and nucleic acids. Life would need to be

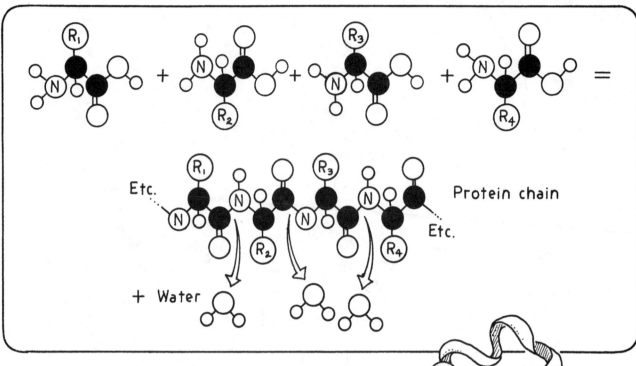

Protein chain

+ Water

resourceful to use the Sun's light to best advantage.

As time passed, the simpler organic molecules began to accumulate in the ancient seas. The "warm little pond" had become a chemical storehouse, a rich broth of organic ingredients. Without free oxygen around to burn them up, these compounds had an extended existence. The stage was set for another level of synthesis.

The next step was a big one: the creation of the very large molecules, the proteins and nucleic acids which are the threads of life. Proteins are chains of hundreds of amino units, linked together in a particular sequence. The nucleic acids are even more complex than the proteins. The nucleic acid DNA, which contains the genetic code of living organisms, is a chain of sugars, phosphates, and organic bases which can have more than a billion links!

It is clear enough what must have happened, even if we don't know precisely how.

To make a protein, amino acids must link up in a particular sequence. Each link is established by the removal of a water molecule. As the chain assembles it twists into a helix, like a telephone cord. At the same time the helix folds up into a crumpled cross-linked shape which is determined by the sequence of amino acids. The shape of the protein decides its role in life. For

Enzyme
lysozyme

example, it is the task of the enzyme lysozyme to help bacteria extract energy from sugar. The little "pocket" in the side of the lysozyme molecule gets a grip on the sugar unit and facilitates its breakdown.

In a living cell the blueprint for an entire repertoire of proteins is encoded in the nucleic acid DNA. Another nucleic acid, the "messenger" RNA, copies the blueprint and carries it to the ribosomes, the part of the cell which actually assembles the proteins from amino acids.

The DNA molecule is the key to life. The molecule has the form of a spiral staircase, or double helix. The rails of the staircase are alternating sugars and phosphates. The steps are pairs of organic bases. The sequence of the base pairs along the double helix is the code that tells the cell how to do its job. The code is amazingly compact. If the biochemical instructions encoded in the DNA of a single human cell were expressed in the notations I used in writing this

Segment of DNA

Building a DNA Double Helix

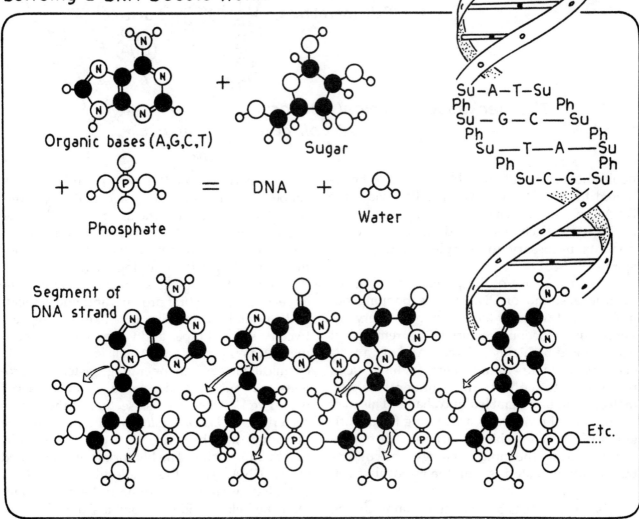

Organic bases (A,G,C,T)   +   Sugar

+   Phosphate   =   DNA   +   Water

Su–A–T–Su
Ph   Ph
Su — G — C — Su
Ph   Ph
Su — T — A — Su
Ph   Ph
Su-C-G–Su

Segment of DNA strand

Etc.

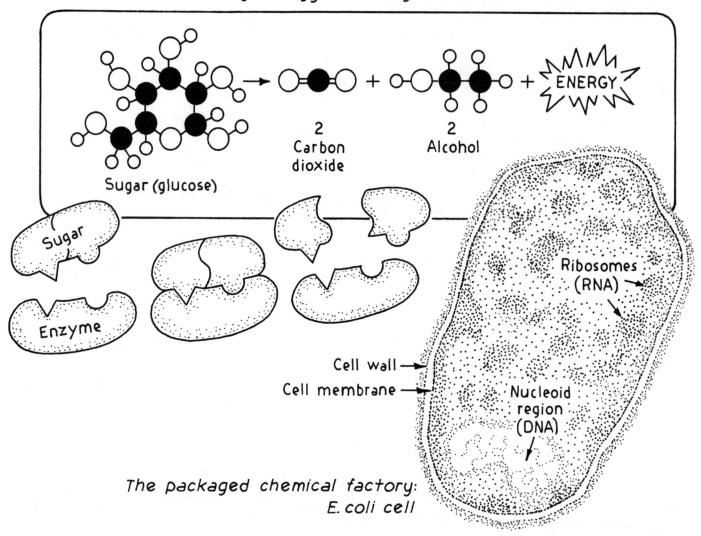

2
Carbon
dioxide

2
Alcohol

ENERGY

Sugar (glucose)

Sugar

Enzyme

Ribosomes
(RNA)

Cell wall →

Cell membrane →

Nucleoid
region
(DNA)

The packaged chemical factory:
E. coli cell

book, thousands of books this size would be required to contain them. When cells divide, the DNA molecule unwinds in a dizzy dance and copies itself. Each side strand attracts the components necessary to complete new double spirals.

As in the case of the protein, creating the links in a strand of DNA means popping out molecules of water. If life began in a watery environment, reactions requiring the removal of water would be highly unlikely. And yet a watery environment would seem to have been necessary to collect and concentrate the molecules to be assembled into chains. This may be the central problem in accounting for the origin of life: how to achieve reactions that require simultaneously the input of energy, the concentration of raw materials, and the removal of water.

Some scientists believe life began in drying ponds, perhaps on the warm flanks of volcanoes. Others have suggested that clays were the catalysts that aided the assembly of the large organic molecules. The truth is, no one knows when or how the first proteins and nucleic acids managed "to get their act together."

It is a further big step that separates the appearance of the first proteins and nucleic acids and the first living cells. It is like the step between a stack of pipes and valves and an operating petroleum refinery. A cell is a "packaged" chemical factory, an operating system of proteins and nucleic acids bound up inside a cell wall. I've sketched a cross-section of the simple bacterium *Escherichia coli*. Within the cell, the ribosomes are the workshops containing RNA that build proteins. The blueprints are contained in the DNA. The DNA molecule in *E. coli*, if unfolded, would be a thousand times longer than the cell is wide. It contains the plans for several thousand proteins. Some of the proteins are used as structural components of the cell, others are enzymes that facilitate the chemical reactions necessary for energy production, growth, and reproduction.

The first cells produced their energy by fermentation, the same process a brewer uses to make alcoholic beverages. Sugar is broken down into alcohol and carbon dioxide, and energy is released from the broken chemical bonds. (There are variations on this theme; see Notes.) The first living cells presumably harvested their sugars from the environment. To be efficient, the fermentation process requires enzymes, the brewer's yeast. The enzymes link up with sugar molecules and speed their disassembly.

The assembly of proteins, nucleic acids, and finally living cells spontaneously from nonliving matter involved a sensitive balance of raw materials, energy, environment and time. Scientists may never see life leap spontaneously from a laboratory flask. Even if they could somehow re-create the physical conditions that prevailed on the Earth 3.8 billion years ago, the one ingredient they can not reproduce is time, geological time, the millions and millions of years that passed on the early Earth like ticks of a clock. Those millions of years allowed ample time for failed experiments, for countless episodes of trial and error.

The story I have told of the origin of life involves matter, energy, and time—and a heaping portion of lucky chance. Those readers who choose to believe for one reason or another that life is more than mere chemistry can take satisfaction in the fact that the story has many gaps. Those who choose to see in those gaps the specific intervention of a supernatural agent or visitors from other stars will do so. Personally, I believe future research will narrow the gaps, and that grounding our religious beliefs in the gaps of science is a risky business. Rather, it seems to me that the more we discover about how life works and how it came to be, the more we are awed by the ultimate mystery of the Universe. "Nothing," said the physicist Michael Faraday, "is too wonderful to be true."

And what of the wild columbine on the mossy outcrop? I know how seeds travel on wind and feather, and I know of the dizzy dance of DNA at the heart of the seed. But the unanticipated flower seems no less a miracle. The proper seat of our reverence resides not in the gaps in our knowledge, but in the Universe itself.

The Window
for Life

The genesis of life on Earth depended on a delicate balance of planet size and distance from the sun.

The Earth was just the right size and just the right distance from the Sun to become a proper "warm little pond" for the genesis of life. How narrow were the limits on size and distance can be appreciated by taking a glance at Venus and Mars, our two nearest neighbors in the Solar System. Apparently the "window for life" was narrow indeed.

By a delightful coincidence, on the very evening I write these lines—Valentine's Day, 1983—I can see Venus and Mars from my window. The two planets are approaching a close conjunction in the western sky. On this particular evening, the Moon briefly adds her slim crescent to the Valentine rendezvous.

I have sketched a view of the western horizon. The last lingering rays of the Sun brush the sky with pale light. Brilliant Venus, ruddy Mars, and the crescent Moon hang above the trees like ornaments. Within the hour they will follow the Sun to a western setting. There are no first magnitude stars in this part of the sky, but the Great

Square of Pegasus, the winged horse of Perseus, provides a suitable backdrop for the planetary conjunction.

There is nothing especially remarkable about a conjunction such as this one. But it is the sort of event, like the surprise visit of a wild columbine, that adds spice to nature's seasonal feast. This Valentine's Day *ménage à trois* of Venus, Artemis, and Mars is a particularly pleasant apparition.

Of course, the coming together of the three celestial bodies in the night sky is more apparent than real. As the drawing below demonstrates, the objects that cling together in the view from my window are actually strung out across 200 million miles of space. The drawing also nicely illustrates the relative distances of the inner planets from the Sun, a matter of some consequence for the origin of life.

At the distance of Venus from the Sun, the Sun's radiation is more than twice as intense as at the distance of the Earth. Mars receives only a third as much sunlight as the Earth. These differences are enough to tip conditions on our

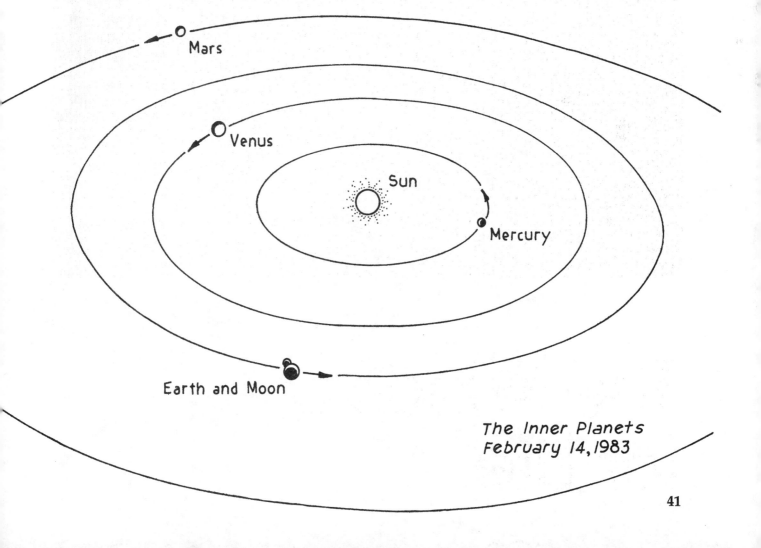

Mars

Venus

Sun

Mercury

Earth and Moon

The Inner Planets February 14, 1983

neighboring planets toward the hellish inferno of Venus or the frozen wastelands of Mars.

It is likely that early in their histories Venus, Mars, and the Earth had roughly similar atmospheres, including similar allotments of water. But today, conditions on Venus and Mars are very different from those which prevail on Earth. Earth's oceans are miles deep and cover three-quarters of the planet. There is no liquid water on Venus or Mars. Carbon dioxide constitutes less than one percent of Earth's atmosphere. The atmospheres of Venus and Mars consist mostly of carbon dioxide. These differences are crucial to the story of life on Earth. The differences are related to the temperatures of the planets, and—ultimately—to the distances of the planets from the Sun. Early surface temperatures on Venus were sufficiently high that most of that planet's water remained in the atmosphere. Liquid water evaporated. Any rain that fell onto the surface sizzled back into the sky. The water vapor in the atmosphere of Venus, together with carbon dioxide, gave rise to a "greenhouse effect."

What is a "greenhouse effect"? The atmosphere of Venus was transparent to sunlight. Sunlight streamed in through the atmosphere and warmed the surface of the planet. The energy was reradiated by the surface as heat. But heat radiation is absorbed by water vapor and carbon dioxide, and so was prevented from escaping back to space. Energy was trapped at the surface and the temperature of the planet soared. The same effect is used in a greenhouse or solar-heated home. Glass, like water vapor and carbon dioxide, lets in sunlight and keeps the heat from getting out.

As the temperature of the planet rose, even more water was evaporated into the atmosphere. A "runaway greenhouse effect" ensued which raised the temperature of the surface of Venus to a scorching 900 degrees F, hot enough to melt lead! No water could exist as liquid at such temperatures.

As time passed, the water vapor in the atmosphere of Venus was disassociated into hydrogen and oxygen by the Sun's ultraviolet radiation or by reaction with hot rock or carbon monoxide. The hydrogen escaped to space, and the oxygen was chemically absorbed at the surface. The result: Venus has very little water in either a liquid or gaseous form. The dense carbon dioxide atmosphere has maintained the "greenhouse." Venus remains an inferno.

By contrast, the Earth was situated at a more comfortable distance from the Sun. The surface temperature of the young Earth soon dropped to a point where water in the atmosphere was able to condense as rain and collect on the surface as liquid. Oceans were in place early in the planet's history.

The oceans had a dramatic effect on the composition of the Earth's primitive atmos-

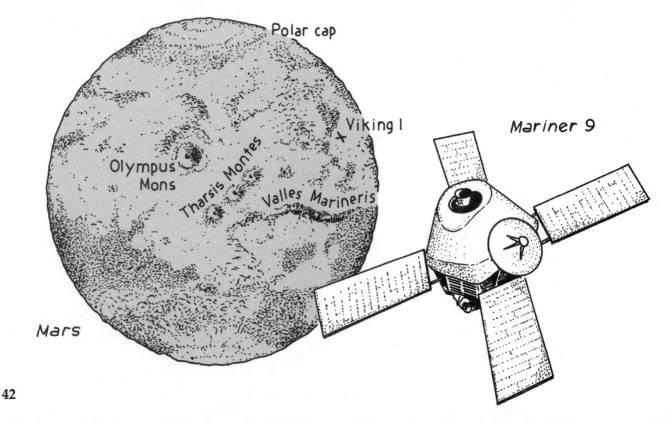

Polar cap

Olympus Mons

Tharsis Montes

Valles Marineris

Viking I

Mariner 9

Mars

phere, which like the atmospheres of Venus and Mars consisted mostly of carbon dioxide. In the presence of liquid water, carbon dioxide combined with dissolved substances to form minerals known as carbonates. Living organisms in the sea used dissolved carbon dioxide to construct carbonate shells. All of these carbonates settled onto the sea floor to form sedimentary rocks.

If you are looking for the water vapor that was in the Earth's original atmosphere, you will find it in the seas. If you are looking for the carbon dioxide, you will find it in the rocks such as limestone and dolomite. The gases that remain in the Earth's atmosphere (at a pressure only a fraction of that of the atmosphere of Venus) do not absorb heat radiation. There is no "glass" for a "greenhouse" on Earth.

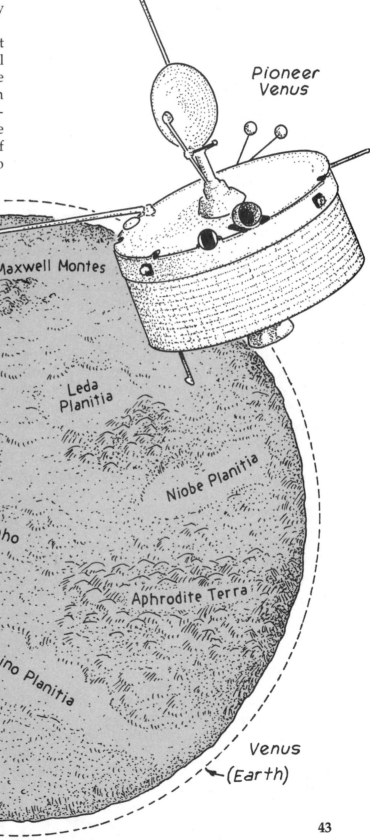

43

Still a different situation prevailed on Mars. At Mars' distance from the Sun, surface temperatures are such that water remains frozen. The great polar ice caps on Mars are partly "dry ice" (frozen carbon dioxide) and partly frozen water. More frozen water may be trapped underground on the red planet. There are no oceans on Mars, no ponds, or streams that might be nurturing environments for life.

Space probe photography has revealed what appear to be water-eroded channels on Mars. The channels suggest that there was liquid water on the planet at some time in its past. Perhaps the channels resulted from the catastrophic volcanic melting of subsurface ice. The channels are a mystery that awaits future exploration.

The lower temperatures on Mars provide one other obstacle to the synthesis of life from inanimate compounds. The tempo of chemical reactions is temperature dependent—the higher the temperature, the more rapidly reactions proceed. The kinds of chemical link-ups that are needed to build large organic molecules would proceed far more slowly on chilly Mars than on Earth.

So like Poppa Bear's porridge, Venus was too hot. Like Momma Bear's porridge, Mars was too cold. Like Baby Bear's porridge, the Earth was just right for water to exist as a liquid, for carbon dioxide to be bound up in rocks, for temperatures to remain moderate, and for life to appear in the "warm little pond."

Some scientists believe that the thermal "window for life" was extremely narrow. A computer simulation by the astronomer Michael Hart suggests that if the Earth had been only five percent closer to the Sun—88 million miles rather than 93 million miles—a "runaway greenhouse effect" would have given our planet a surface like the blazing hell of Venus. Only one percent further from the Sun—94 million miles—and runaway glaciation would have resulted in an icebound planet. If Hart's calculations are correct, life may exist on Earth by the slimmest of chances.

The size and density of the planets also affect conditions for life in several important ways. The inner planets have similar densities and probably are made of much the same stuff (the density of Mars is somewhat less than Mercury, Venus, and the Earth). The relative sizes of Mars and Venus are shown to the same scale on the previous pages. An outline of the earth is added for reference. Venus and the Earth are almost perfect twins in size and density.

Size and density affect the strength of gravity at a planet's surface. A greater surface gravity gives the planet a stronger grip on its atmosphere. The Earth's moon is too small to hold any atmosphere at all. Mercury has a greater density than the Moon and a somewhat larger size, but high temperatures on Mercury encourage the escape of atmospheric gases to space. Any gases that were released on Mercury or the Moon during early cratering and volcanism have long since been lost. Mars also had a hard time holding on to its atmosphere. The atmospheric pressure on Mars is only 1/150th that of the Earth.

But Mars may never have had an atmosphere as dense as the Earth's. A smaller planet has more surface compared to its volume. This allows for a more rapid cooling of the interior. As a result, Mars quickly developed a thick stable crust. Mars has probably experienced less volcanic activity than the Earth or Venus and expelled less gas from the interior.

In contrast to the other planets of the inner Solar System, the Earth was just the right size and just the right distance from the Sun to make and hold an atmosphere with a composition and density perfectly suited for the genesis of life.

No one seriously expects that life could survive on the hellish surface of Venus. Veneras 9 and 10, the two massive crafts that the Russians successfully landed on the surface of Venus in 1975, lasted only an hour in the formidable heat before falling silent. Mars, on the other hand, offers intriguing possibilities for the discovery of life. Surface temperatures on Mars are colder than on Earth and range more widely, but they do not rule out the survival of hardy terrestrial organisms. Certain organisms, after all, do quite well in the inhospitable environments of Antarctica. Channels on the surface of Mars imply there was liquid water at some time in the past. If life exists anywhere else in the Solar System it will likely be on Mars.

Ever since the discovery of the so-called "canals" of Mars by Secchi in 1869, the idea of life on the red planet has had a powerful hold on the human imagination. The "canals," of course, turned out to exist only in the eye of the beholder. No contemporary scientist expects that higher life forms make their home on the dusty

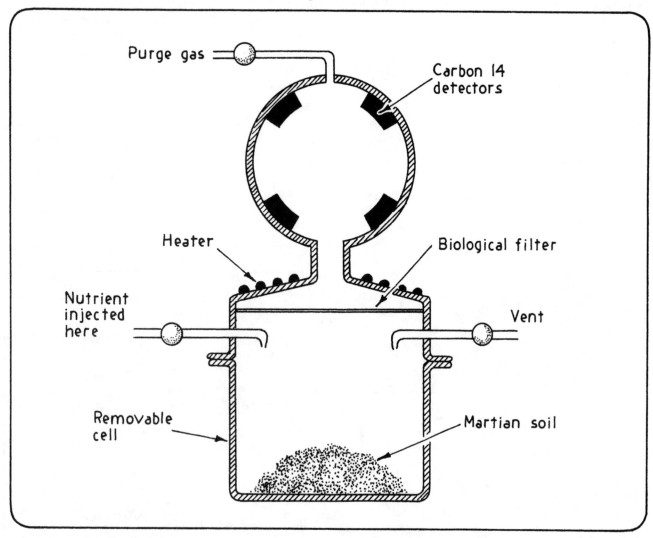

Purge gas

Carbon 14 detectors

Heater

Biological filter

Nutrient injected here

Vent

Removable cell

Martian soil

deserts of Mars. But microorganisms in the Martian soil seem a reasonable possibility, and the search for life was a central motive for the two Viking lander missions to the red planet.

The Viking landers carried three separate experiments to search for extraterrestrial life. All three were based on the premise that living organisms would alter their environment in some way. On this page I have illustrated the apparatus used in one of these experiments, the so-called "labeled release experiment." A sample of Martian soil was placed in the cell of the apparatus by Viking's mechanical arm. The cell was sealed to the apparatus and the soil was moistened with a nutrient broth rich in vitamins and amino acids ("chicken soup," the scientists called it). If microorganisms were present in the Martian soil they would presumably "eat" and metabolize the broth, releasing gases. The "chicken soup" was laced with radioactive car-

bon 14. After letting the Martian "bugs" feast for a few days, radioactivity detectors looked for metabolized gas.

There was always the possibility that ordinary chemicals in the Martian soil might react with the nutrient and release gas. Many chemicals "fizz" when moistened. To eliminate this possibility the apparatus was equipped with a heater so that soil samples could be sterilized. Heating would supposedly kill the microorganisms without affecting chemicals. The apparatus was tested on Earth soil before being sent to Mars. The detectors gave positive results with unsterilized Earth soil, and negative results with heated samples. Earth bugs make gas! Earth bugs are killed by heat! The device was also tested on samples of lunar soil. The detectors gave negative results with both sterilized and unsterilized lunar samples, as one would expect of the soil of the lifeless Moon.

Scientists therefore had every expectation that if Mars harbored life, Viking would find it. The experiment was run several times on the red planet at both landing sites, and always with the same results. Unsterilized soil released gas when moistened with "chicken soup," and sterilized soil did not! It looked as if something was "eating the soup." The evidence for life on Mars seemed tantalizingly real.

Unfortunately, two other experiments designed to test for photosynthesis and respiration gave totally negative results. A purely chemical explanation for the results of the labeled release experiments might be possible. At this point,

most scientists remain skeptical about life on Mars, but the case must be said to remain open.

Vikings 1 and 2 landed on the surface of Mars in July and September of 1976. They radioed back magnificent photographs of the planet's rocky surface, but no little green Martians cavorted in front of the cameras. A central task of the two crafts was to discover just how wide or narrow is the "window for life" in the Solar System. If life exists on Mars, then the window may be wide enough that we can reasonably expect to find the Universe teeming with life. If Mars is sterile—as most scientists still expect, in spite of the labeled release

Viking Lander

experiment—then the "window for life" is narrowed, and life on Earth looks more and more like a lucky chance.

It is easy to ask "What if?" What if the Earth had been 20 percent smaller, or 10 percent farther from the Sun? What if the "warm little pond" had frozen over or boiled away? What if the atmosphere had been too rarefied to concentrate life-creating compounds, or so thick that little sunlight reached the surface? What if? There are a thousand what ifs.

We know that even the rarefied space between the stars is rich in organic compounds. It may be that the tendency of matter to assemble into complex and resourceful forms is so powerful that life is inevitable under a wide variety of circumstances. If that is so, then the "window for life" is thrown wide open. On the other hand, our one certain example of life in the Universe may have been the lucky and unique product of a thousand chance factors.

Several days have passed since I began this essay. Mars and Venus have nudged even closer in the western sky. I know that their proximity is an illusion of my limited perspective. The two planets are actually separated in the Solar System by a yawning gulf.

In the same way, I suspect the apparent narrowness of the "window for life" is an illusion of our limited perspective, and that future discoveries will show that life is more resourceful than all the obstacles which circumstance throws up against it. Life may be able to get a start in a wide range of environments. And once initiated, life has a way of modifying its environment to its own liking.

I would be very surprised if we were alone.

An Uncertain
Star

The Sun's radiation may not be as constant as was once supposed.

Stonehenge, the Salisbury Plain, 75 miles southwest of London, England. A midsummer night. The air is chill, magical, mysterious. The Great Bear—we know him better as the Big Dipper—stalks the starry sky.

Summer nights are short at these latitudes. Only a few hours of true darkness separate the twilight from the dawn. But the darkness is cold. The darkness is fearful. Those few hours are enough to confirm our dependence on the Sun.

We wait in the circle of standing stones watching the horizon in the northeast. The sky grows light. The limb of the Sun's disk appears, in ripples of refraction, precisely where we expected. The blazing red globe slices upward at an angle, slowly disentangling itself from the horizon. Then for an instant it is poised on the "heel stone," its triumphant arrival bracketed by the posts of the "gate."

The axis of the prehistoric circle at Stonehenge, with "gate" and "heel stone," is aligned with the sunrise on the summer solstice. The summer solstice is that day of the year when the Sun reaches its northernmost excursion in the sky. In northern latitudes the solstice brings warm days and short nights.

Stonehenge tends to be crowded these days on the morning of the solstice. Tourists and latter-day druids and peripatetic astronomers assemble to share an ancient ritual. Only about half of the original stones are still in place. But the Sun comes up and dances on the "heel stone" just as it did 4700 years ago when the building of Stonehenge commenced, and as it did forty generations and a thousand years later when construction of the great monument was finally completed. Stonehenge was then and is now a place to rejoice in the constancy of the Sun.

Stonehenge is a more sophisticated solar observatory than the crude arrangement of stone slabs we visited at Fajada Butte in New Mexico. But the purpose—or at least part of the purpose—was the same. Like the Anasazi of Chaco Canyon, the builders of Stonehenge were an agricultural people. They depended for their livelihood on the generous radiation of our yellow star. They marked with stone the passage of that star across the sky.

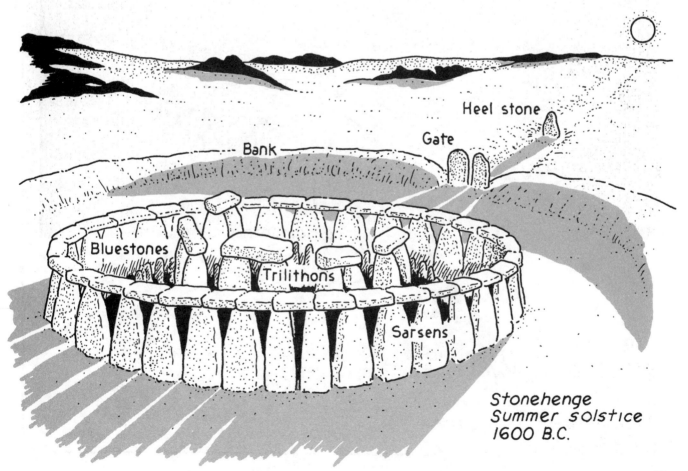

Heel stone

Gate

Bank

Bluestones

Trilithons

Sarsens

Stonehenge
Summer solstice
1600 B.C.

We are today better insulated by technology from the vagaries of climate and seasonal change than were the Anasazi of New Mexico or the builders of Stonehenge. But we understand with no less certainty that life is the gift of the Sun. We too value the Sun's constancy.

When I first studied astronomy 25 years ago, the constancy of the Sun's radiation was as firmly held as any other "fact" in the textbook. We memorized the so-called "solar constant," the rate at which solar energy reaches the Earth's surface (1.4 kilowatts per square meter). The Sun, we were told, is a rock-steady star. That steadiness, it seemed, is a corollary of the orderly evolution of life on Earth.

In recent years the Sun's steadiness has been called into question. But before we consider this new development, it would be well to survey some of the more familiar aspects of our yellow star.

To the naked eye the Sun is a flawless globe. The invention of the telescope in the 17th century revealed disturbing dark spots on the face of the Sun. The spots were only a hint of what was to come. Modern instruments reveal the Sun's surface as a turbulent ocean of gas ceaselessly stirred from within and lashed by stupendous magnetic storms.

Time-lapsed color movies of great solar storms are among the most impressive artifacts of science I have experienced. No justice can be done to the drama of a solar storm with a static black-and-white illustration. Nevertheless, I have sketched here a typical solar flare against the backdrop of the Sun's churning surface.

The Sun's surface is mottled by granules of hot gas that seem to rise from the interior like bubbles in a vigorously boiling liquid. Beneath each granule is a convection cell, a part of the Sun's outer layers where hot gas rises and cooler gas sinks, bringing energy to the surface, stirring the Sun's bulk. Flickering columns of flame, called *spicules*, dance above the granulated surface.

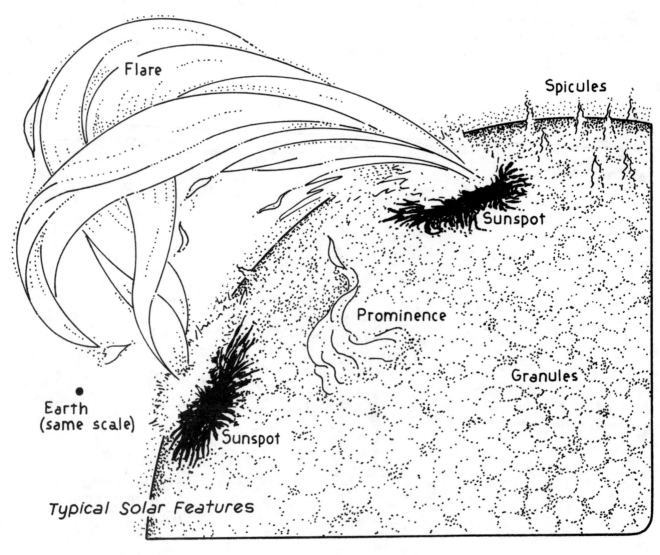

Flare

Spicules

Sunspot

Prominence

Granules

Earth
(same scale)

Sunspot

Typical Solar Features

The famous sunspots are actually storms of luminous gas. They appear dark only by contrast with the hotter surrounding surface. The spots are controlled by the Sun's twisted, kinked, looped, and knotted magnetic field. From the regions of the spots, tongues of flame, called *prominences*, soar outward along lines of magnetic force. They hurl great sheets of burning gas into space. The most energetic of these prominences, and the turbulent sunspot sites that anchor them, are called *flares*. Flares can lash outward for a distance equal to half the Sun's diameter. In my drawing I have added an out-of-place Earth to provide a sense of scale. Solar firestorms make the Earth look puny indeed!

The Sun's visible radiation is not greatly affected by solar flares, but during a flare, x-ray radiation can increase a hundredfold. Flares also fling into space streams of charged particles—the solar wind—that wash across the Earth as if it were a grain of beach sand facing the fury of the sea. Even at a distance of 93 million miles the consequences for life are potentially dangerous (see Chapter 13).

Photographs of the Sun made with x-rays or ultraviolet light give the lie to the eye's impression of a passive stolid object. They show instead a swirling, explosive ball of luminous gas in constant turmoil. All of this energy has its origin deep in the Sun's core where hydrogen is fused into helium, and mass is turned directly into radiant energy. This energy moves outward through progressively cooler layers of hydrogen and helium. For most of this journey, energy transport is radiative, by the successive absorption and reradiation of energy. Near the surface, energy transport becomes convective, carried upward by a circulation of the Sun's substance. The visible light of the Sun is radiated at the photosphere. This is the layer of luminous gas that gives the Sun the appearance of a sharply defined globe. The Sun does not have a true surface—if you could survive the 5700 degree (K) temperature, you could fall right through the photosphere into the Sun's interior.

Above the photosphere is another layer of gas, the chromosphere, a kind of froth that glows with the pinkish light of hydrogen. Then comes the corona, the thin hot outer atmosphere of the Sun. The corona is visible during an eclipse of the Sun as a brilliant outward-streaming halo of light. At other times the corona's delicate structure of wisps and spikes is

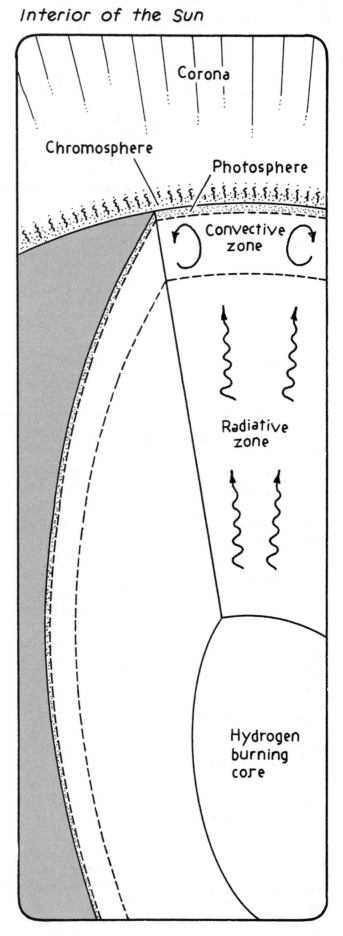

Interior of the Sun

Corona

Chromosphere

Photosphere

Convective zone

Radiative zone

Hydrogen burning core

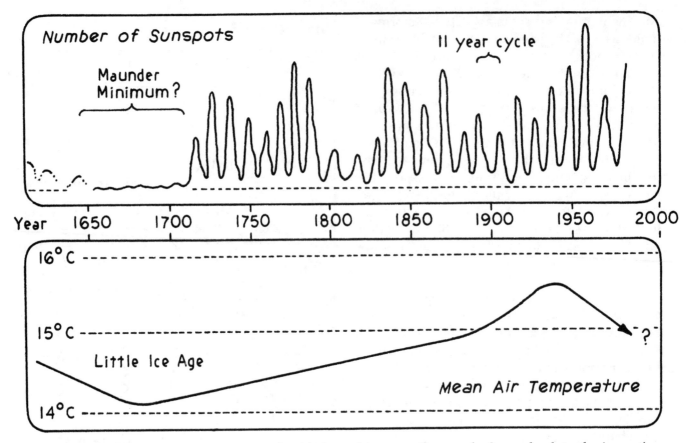

Number of Sunspots

Maunder Minimum?

11 year cycle

Year 1650   1700   1750   1800   1850   1900   1950   2000

16°C

15°C

Little Ice Age

Mean Air Temperature

14°C

washed out by the overwhelming light of the photosphere.

In recent years the corona has been studied from high-altitude observatories and Earth-orbiting satellites with instruments that create a kind of artificial eclipse. The corona is a highly rarefied gas, but fantastically hot, far hotter than the Sun's apparent surface. It is probably heated by shock waves from the violent subsurface convection region. The corona, like all the Sun's outer layers, is stirred and given shape by the Sun's powerful magnetic field.

It would perhaps not be surprising to find that so violent a furnace varies a bit in its output of energy.

Recent suspicions of the Sun's inconstancy began with solar physicist John Eddy's rediscovery of the so-called "Maunder minimum." Eddy confirmed from historical records something first noted in the late 19th century by the German Gustav Sporer and the Englishman Walter Maunder: sunspots seem to have been virtually absent from the Sun's surface during the period from 1645 to 1715.

It has long been recognized that the number of spots on the Sun's face goes through an 11-year cycle. The cycle is believed to be controlled by the Sun's magnetic field. The 11-year cycle is

shown on the graph above, back to the invention of the telescope in the early 17th century. The absence of sunspots during the Maunder minimum is notable. Before the 17th century the only available sunspot data are occasional naked-eye observations by Chinese astronomers (see Note).

The Maunder minimum roughly coincides with a period in European climatic history known as "the Little Ice Age." From about 1400 to 1850, and climaxing during the Maunder minimum, Europe's climate was significantly colder than during the periods before or after. Ice festivals were held on the frozen Thames River in London, and skaters frolicked all winter on Dutch canals. Crops failed across Europe, and glaciers slid down their valleys in the Swiss Alps like mercury in a thermometer.

Eddy tentatively suggested that there was a relationship between sunspot activity and climate, or rather that the absence of sunspots during the Maunder minimum was related to the same aberration of the Sun's activity that produced the chilling temperatures. It would not take a major change in the Sun's activity to have a chilling effect on Earth. A drop in the Sun's luminosity by only one percent would be enough to trigger a Little Ice Age.

Since Eddy announced his findings in 1976, the record of solar activity has been extended back thousands of years. The record has been inferred from historical accounts of naked-eye sunspots, observations of the solar corona during eclipses, and displays of the auroral lights. The Sun's past activity can also be deduced from the amount of carbon 14 in the rings of very old trees. Carbon 14 is produced in the Earth's atmosphere when cosmic rays from space strike atoms of nitrogen. Solar storms can partially block cosmic rays from reaching the Earth. Growing trees incorporate carbon from the atmosphere into their tissue and so keep a kind of running record of activity on the Sun.

The curve of solar activity based on all of this data shows intriguing similarities to the curve of climate. Climate is an incredibly complex system, and the role of solar inconstancy is hotly debated. But this much is clear: the Sun may not burn as steadily as we once believed, and the effect on Earth might not be insignificant.

Another hint of a possible inconstancy in the Sun is the "case of the missing neutrinos."

Deep in the Sun's core, hydrogen is transformed into helium by nuclear fusion. The most obvious product of that process is the radiant energy which sustains life on Earth. But if you look back to the diagram on page 5 you will see that there is one other product of fusion, something I didn't mention in the earlier chapter, the tiny elusive particles called neutrinos.

Neutrinos have been a part of physics since they were introduced as hypothetical entities in 1931 by the physicist Enrico Fermi. They were detected experimentally in 1956 with an elaborate apparatus installed next to a major nuclear reactor. Neutrinos have no charge and seemingly no mass. They come about as close to being nothing as anything can be and still be real. They do carry energy and they do have measurable properties. And they are produced in prodigious numbers at the heart of the Sun.

Neutrinos have one property which attracts our interest here—they almost never interact with ordinary matter. Shoot a neutrino at a stone wall and it will pass through. Shoot a neutrino into the ground and the odds are nearly perfect that it will arrive at Australia unimpeded!

The copius flux of ghostly neutrinos created at the Sun's core should pass up through overlying layers of the Sun's bulk and out into space with negligible diminishment. It takes millions of years for the Sun's radiant energy to make its way from the core to the surface. But in two seconds neutrinos are out and away. Eight minutes later a tiny fraction of this steam intercepts the Earth. Most of the neutrinos pass through the Earth as if it weren't there.

If physicists have reasoned rightly about what is happening in the Sun, then 80 octillion neutrinos (8 followed by 28 zeros) stream through the Earth each second. A hundred trillion neutrinos pass through your body every second, indoors or out, day or night! Surely few statements made by physicists could be more bizarre.

Raymond Davis of the Brookhaven National Laboratory has designed an experiment to detect the solar neutrinos, assuming of course that they are there. His "telescope" is a huge tank of the cleaning fluid tetrachloroethylene, a mile underground in a gold mine near Lead, South Dakota.

Of course, the very property of neutrinos that enables most of them to soar unsnagged from the Sun's core to the Earth also makes them devilishly difficult for Davis to catch. Davis works with the remote but calculable probability that an occasional neutrino will react with a chlorine atom in his tank. The hoped-for reaction results in the transformation of a chlorine atom into argon-37, a radioactive form of argon which should be fairly easy to detect. The tank is underground and—when operating—submerged in water to shield it from cosmic rays and neutrons from natural radioactivity in the surrounding rocks. Davis wants nothing but solar neutrinos to stop in his tank. (See Note.)

The experiment has been running for 15 years. It has been carefully examined and reexamined for possible flaws. The results are unsettling for physicists.

The best theories for what's going on at the Sun's core predict that Davis's apparatus should snare about one neutrino per day. The actual rate of capture is only about a third of that.

You may have already decided that this neutrino business is "off the wall," and that trying to detect a couple of argon atoms in 100,000 gallons of cleaning fluid makes finding the proverbial needle-in-a-haystack look easy. But physicists have a reasonable confidence in each separate piece of the story I have told and are puzzled that they don't add up.

It may be that our theories about what makes the Sun shine are wrong. It may be that Davis's neutrino detector contains a hidden defect. Or it may be that the nuclear furnace at the center of the Sun is not burning as furiously as we think.

Predictions of the number of neutrinos produced at the Sun's core are based on the amount of heat and light which is radiated at the Sun's surface. The radiant energy of sunshine requires millions of years to "gurgle up" from the core. How brightly the Sun is shining today really tells us what was going on "downstairs" in the solar basement two million years ago. Neutrinos, on the other hand, sprint from the core of the Sun to the Earth in eight minutes. They zip out through half a million miles of the Sun's substance and across 93 million miles of space like eager messengers. They carry information about how vigorously the furnace at the center of the

Sun is burning this very day. If the rate of nuclear burning deduced from sunlight does not match up with the rate deduced from captured solar neutrinos, it may mean that the solar furnace has cooled down. If that is so, then sometime in the next few million years we may be in for a chill.

This last explanation for the low number of neutrinos caught by Davis with his "neutrino telescope" may be unlikely, but it has been seriously entertained. The calculations and the experiment are being refined. Stay tuned for the outcome.

In the summers, I live and work in the west of Ireland (you have surely guessed an Irish connection by now). On the hill above my house is a megalithic ruin known locally as the Giant's Bed. It is a single-chambered tomb fashioned from slabs of sandstone at about the time construction at Stonehenge was begun.

Neutrino Telescope
Homestake Mine
Lead, South Dakota

The Giant's Bed — County Kerry, Ireland

I was up there recently, confirming for myself the east-west orientation of the tomb. Five thousand years ago, a great chieftain was laid to rest, with his head and his feet to the rising and setting of the equinoctal Sun. The tomb was sealed and covered over with a mound of earth. The chieftain slept for the ages.

The earthen cover has blown away, the stones have fallen askew, the bones of the chieftain have turned to dust. But twice each year on the equinoxes, the Sun still rises at the head of the grave and sets at the foot, and shines—we would like to assume—with the same steady light.

Surely, for the tribal chieftain of ancient Ireland, nothing in this world seemed so immune to mortal vagaries as the Sun. In the alignment of his tomb, he perhaps sought to associate himself with the Sun's eternal constancy. Our perspective is more skeptical. In recent years the Sun has provided disturbing hints that it may not be so constant after all. We may yet learn to live with an uncertain star.

Canada Mayflower

Harvesting the Sun

Sunlight drives all biology. Early in its history, life learned to use sunlight to make food. The side effects on the terrestrial environment were considerable.

There comes a moment in New England woodlands in the spring when up through last season's brown oak leaves and the matted needles of the solitary white pine comes the first green. Like a carpet unrolled overnight, suddenly the greedy leaves of the Canada mayflower are everywhere.

The Canada mayflower—sometimes called wild-lily-of-the-valley—insinuates a web of runners beneath the leaf litter, colonizing dark continents of the forest floor. From this hidden infrastructure of communication and transport, the plant throws up a thousand green leaves, one to the right and one to the left, like supplicant hands. In a few weeks there will be tiny white flowers, and the plant will be busy with arrangements for the next generation. But for the moment, the business is pure energy. Capturing sunlight. Soaking up the rays.

There is little enough sunlight on the floor

of the oak woodlands. The Canada mayflower must aggressively court its share. Hence the number of plants, the broad elf-sized forest of green leaves that so enchants the walker in the spring woods.

Reaching for the sun, each leaf shoulders aside the detritus of last year's decay. No law of physics is more basic than the law of entropy, the tendency of the universe to move toward disorder and death. But life bucks that tide, using available free energy when and where it can find it to build elaborate and elegant molecular mechanisms for staying alive, for metabolism, motility, and reproduction. Staying alive has a color and that color is green.

All life on Earth (or almost all) fuels itself on sunlight. Early in the game, life found a molecule—chlorophyll—that was particularly efficient at trapping and temporarily storing energy from visible light. Chlorophyll does not equally absorb all parts of the solar rainbow; if it did, plants would be black. The molecule absorbs red and blue light selectively, rejecting green. Our planet is green because of that part of

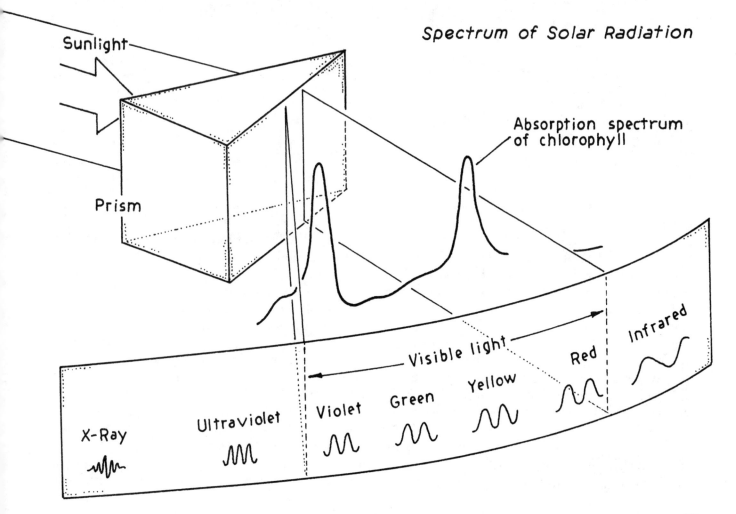

Spectrum of Solar Radiation

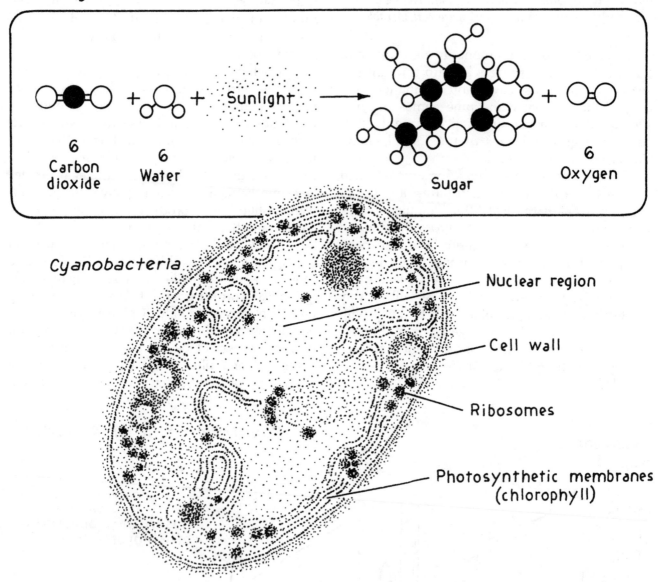

6
Carbon
dioxide

6
Water

Sunlight

6
Sugar

6
Oxygen

Cyanobacteria

Nuclear region

Cell wall

Ribosomes

Photosynthetic membranes
(chlorophyll)

the solar feast which life sends back from the table.

The first primitive self-replicating cells that appeared almost miraculously three and a half billion years ago in the "warm little pond" required fuel. Those first microorganisms obtained their energy by fermentation, by breaking sugars down into alcohol and carbon dioxide (or similar products) and utilizing the energy which was stored in the broken chemical bonds. The sugars were scavenged from the environment. The Earth at that time was bathed with ultraviolet light and crackled with electrical energy. Lots of big syrupy molecules were cooked up in the steamy seas. Living cells devoured them.

But life was living on borrowed time. The

sugars in the seas were a limited resource. Life was a dervish driven by the dizzy compulsive dance of the DNA. Organisms proliferated faster than the food supply. A great dying-out was in the offing.

And then came one of the peak moments in the history of life on Earth, the invention of photosynthesis. It is tempting to say "just in the nick of time," but I have the sense that it may have been inevitable. Life was too clever not to solve the problem of its food supply and figure out a way to use the energy of sunlight directly.

The first photosynthesizing organisms may have resembled the single-celled cyanobacteria, traditionally called blue-green algae. Special membranes in these cells contain chlorophyll which absorbs sunlight. The energy is passed along to the chemical factory which is the cell itself. The process is complicated, but the net effect of photosynthesis is to assemble hydrogen atoms and carbon dioxide into sugar.

The first photosynthesizers probably got their hydrogen from volcanically produced hydrogen sulfide. The waste product of the reaction was sulfur. A breakthrough came with a process for getting the hydrogen from water. Water was an inexhaustible and ubiquitous resource.

Photosynthesis involves many steps, but the bottom line is that six molecules of carbon dioxide and six of water are charged with sunlight to yield sugar and oxygen. The sugars are then available for fermentation, tidy packets of ready energy. Life had learned how to create its own food supply. No more scavenging a living from the land. Life had gone on to subsistence farming.

But this successful innovation was not an unclouded good. The solution of one problem—the food crisis—exposed life to two new dangers— ultraviolet light and oxygen.

Photosynthesizing bacteria require visible light, and early organisms quickly evolved ways to sense the light and move toward it. Free floaters gave way to swimmers. Out from dark nooks and crannies they came, wiggling microscopic fins and tails, to bask in the sunlight.

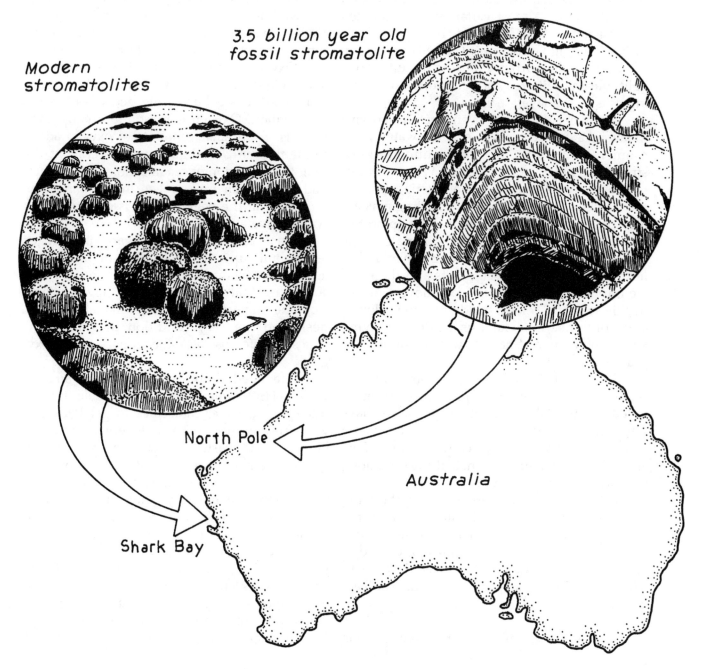

Modern stromatolites

3.5 billion year old fossil stromatolite

North Pole

Shark Bay

Australia

At the surface of the pond, the photosynthesizers found the sunlight they needed to make food. But they were also exposed to the Sun's short wavelength ultraviolet light. Ultraviolet radiation breaks up nucleic acids and proteins. Ultraviolet light can be fatal.

Various clever stratagems were employed by early cells to avoid the danger of ultraviolet light. Organisms may have lived in solutions rich in salts which absorb ultraviolet rays but pass visible light, or beneath transparent grains of sand which do the same thing—a kind of sun glasses, if you please. Or they may have developed pigments in their outer walls that harmlessly absorbed ultraviolet rays, protecting the guts of the cell and the delicate DNA—a kind of suntanning. Almost certainly they adopted the macabre stratagem of living in self-protecting matlike colonies. The cells on the top of the mat died from exposure to ultraviolet light, but their littered corpses protected the living cells in the layers below.

Matlike colonies of cyanobacteria survive today in a few protected shallow-water environments. They build layered hummocklike structures called stromatolites. When a wave washes across a mat of cyanobacteria, the cells trap mineral grains in the water. The grains are incorporated into the living mat and help protect the cells from ultraviolet light. Ultimately, the cells die or move upward seeking sunlight. The next wave brings new sediments and these too are trapped. Again the survivors move upward. In this way the colony builds up stony domes and columns. One famous colony of living stromatolites is at Shark Bay in western Australia.

Not far from the living stromatolites at Shark Bay, at a hot desert place incongruously called North Pole, Australian geologists have found what appear to be fossil stromatolites 3½ billion years old. If the structures at North Pole are indeed stromatolites, then they are among the earliest evidences for life on Earth. They also suggest that bacteria were photosynthesizing food within a few hundred million years of the origin of life on Earth. By colonizing suitable environments, life had learned to take the good in sunlight and avoid the bad, to bask in the sun with only mild attrition, to "have its cake and eat it too."

The waste product of photosynthesis is oxygen. As life went about making sugar with sunlight, free oxygen built up in the environment.

Eventually, the oxygen would solve once and for all the problem of the ultraviolet. An ordinary oxygen molecule has two atoms of oxygen. In the upper atmosphere the energy of ultraviolet rays can break up molecules of ordinary oxygen, which then recombine into a rarer form—ozone—with three atoms. Ozone absorbs ultraviolet radiation effectively. An ozone layer in the Earth's atmosphere now provides an efficient screen against the Sun's deadlier ultraviolet light.

So it would seem like oxygen was a blessed side effect of photosynthesis. Not so! Eventually life would come to rely on oxygen in several ways, indeed to thrive on it. But at the beginning, oxygen was a clear and present danger.

Oxygen is promiscuous. It has an irresistible tendency to combine with almost everything. Iron rusts, wood burns, life decays—all forms of oxidation. Oxygen is poison to organic compounds. Even small amounts of oxygen added to the flask in Urey-Miller type experiments inhibit the formation of organic molecules. If there had been much free oxygen in the Earth's atmosphere at the beginning, it is doubtful if life would have evolved at all.

Once photosynthesis began and oxygen levels built up, the probability of life arising again spontaneously dropped to zero. It is not surprising that we all seem to have evolved from the same root stock. Early in its history life destroyed the very condition that made the genesis of life possible—the absence of free oxygen.

Life needed hydrogen to make sugars. Water—H_2O—was a readily available source. But splitting water molecules to get hydrogen was a risky business. The leftover oxygen was a kind of toxic waste. Oxygen oxidizes, oxygen burns. Life may have saved itself from burning up by rusting its environment.

Among the more intriguing ancient rocks of the Earth's crust are the so-called banded-iron formations. These metamorphosed and twisted rocks had their origin between 3 and 2 billion years ago. They are characterized by alternating layers of dark and light iron-rich minerals. The dark bands are oxygen-poor minerals, and the light bands are oxygen-rich rustlike compounds. The formations are widespread on Earth and

Banded Iron Formation

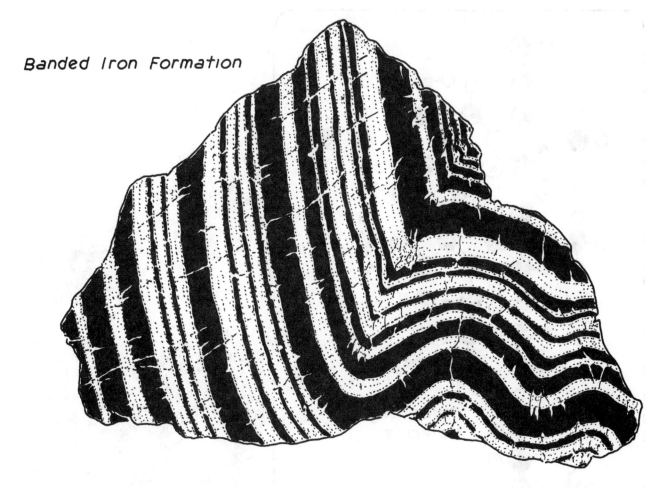

valued as iron ore. The origin of the rocks was long a geological puzzle.

Iron is one of the more common elements on Earth and was undoubtedly common in the early environment dissolved in sea water. When photosynthesizing organisms began producing oxygen, the oxygen combined with dissolved iron to form iron oxides (rust). The oxides precipitated out of solution and with silicon compounds accumulated on the ocean floor. Variations in the availability of iron or oxygen accounted for the finely-layered character of the sediments. The sediments were consolidated into rock and later folded up onto the continents to become the banded-iron formations.

For hundreds of millions of years, early photosynthesizing organisms lived in a kind of precarious equilibrium with dissolved iron. The iron took up the toxic waste of photosynthesis and turned it into harmless rust. Finally, about 2 billion years before the present, the oceans had been swept free of dissolved iron and other oxygen-accepting elements. The oceans of the Earth had rusted! Once again, oxygen levels in the atmosphere began to rise dangerously. But

life had used the interval of safety to learn how to tolerate oxygen and turn it to good use. The toxic waste crisis had passed. But that is a story for the next chapter.

With the invention of photosynthesis life learned how to plug into a star. The battle against entropy had been won. From that time forward, order on Earth could proceed at the expense of a greater disorder at the core of the Sun. The universe as a whole continued to run down, as it must, but on the surface of the Earth there spread a film of highly ordered matter of marvelous complexity and resourcefulness.

The organisms that ruled the Earth 2 billion years ago were no more advanced than the scum that lives on your shower curtain, but that scum had learned how to build sugars with sunlight and how to cope with several of the more insidious perils that have threatened life on this planet.

The tree of life has many branches. The cyanobacteria represent one spectacularly successful early line based on the light-trapping molecule chlorophyll. Living in colonial mats

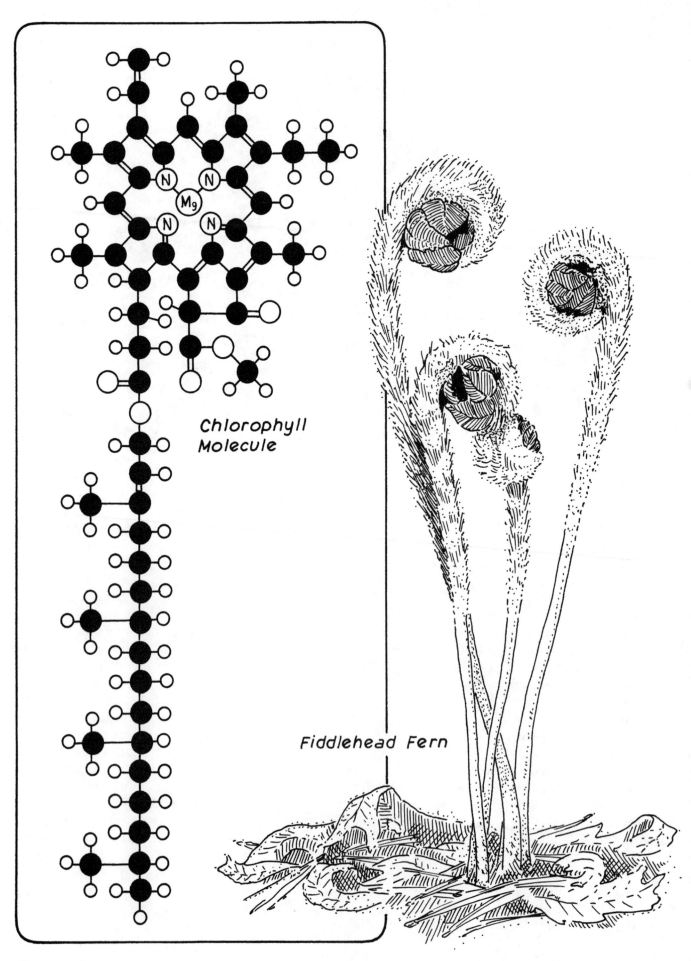

Chlorophyll
Molecule

Fiddlehead Fern

62

and scrambling for sunlight, the cyanobacteria built foliated towers called stromatolites that remain the most durable evidence for early life in the geological record.

Later on the cyanobacteria contributed their sunlight-trapping talents to the first true plant cells. Animals developed along a different branch of the evolutionary tree, and it now seems likely that you and I had no photosynthesizers among our ancestors. But the tree of life is a tree of interdependence. Without plants, animals would not survive. Plants are our food-producing link to our yellow star.

Plants have devised a bag of tricks for grabbing sunlight. For those of us who live in northern deciduous woodlands, those tricks are a recurring source of seasonal pleasure. I have chosen to end this ode to photosynthesis with another of my favorite plants of spring, the cinnamon fern. Each year there is a certain day in April when the Sun burns with a fierce new heat, and the ferns unroll their fiddleheads in sandy soil near the wooded swamp. Like croziers they come up, like cudgels, like Irish shillelaghs, shaking their tight little fists at the winter past. Then with a confident flare the cinnamon fern unfurls its broad sails of chlorophyll, drinking up the Sun's red and blue light and leaving the green for the season.

The Breath of Life

Increasing levels of oxygen in the atmosphere posed a temporary crisis for life. With respiration, life found a way to turn a poison to good advantage.

If the Earth were not tipped on its axis there would be no seasons. There would be climates, certainly—warm equatorial regions, temperate mid-latitudes and frigid poles—but no dramatic annual variations. No winter, no spring, no summer, no autumn. No Canada mayflowers pushing up through winter's leaf litter, transforming the forest floor from brown to green. No fiddlehead ferns unrolling fine-fingered fronds to catch the sun.

Spring provides a kind of annual recapitulation of the evolution of life on Earth, an opportunity to celebrate the greening of the Earth three and a half billion years ago by the photosynthesizing bacteria. All life depends on the photosynthesizers. The creatures that do not have the ability to make food with sunlight—the fungi, the animals, you and I—are inevitably part of food chains with plants, the sun-trappers, at the base. As spring unrolls its green carpet, the non-photosynthesizers get moving too. From rock-hard seed cases, from underground burrows, from behind the bark of trees they come, to eat plants, or to eat the creatures that eat plants, or to eat the creatures that eat the creatures that . . .

Photosynthesis made life self-sufficient. But it also created the first toxic waste crisis. For a while, the deadly oxygen which was a by-product of photosynthesis was taken up by iron and other oxygen-grabbing elements in the environment. The poison was harmlessly incorporated into the rocky crust of the Earth. But eventually this "toxic waste dump" for oxygen was exhausted, and the atmosphere began to change. Just how drastically the Earth's atmosphere has been modified by photosynthesis can be seen by comparing the atmospheres of Earth, Venus, and Mars.

It is likely that the Earth's early atmosphere was similar to the present atmospheres of our

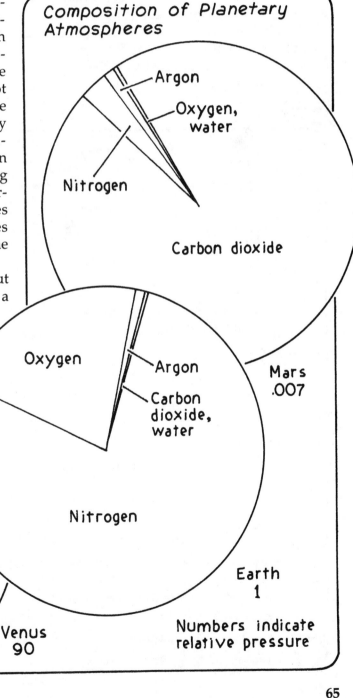

neighboring planets. That situation no longer prevails. The atmospheres of Venus and Mars have large components of carbon dioxide and almost no oxygen. Earth's atmosphere is one-fifth oxygen and contains almost no carbon dioxide. What accounts for the difference?

The absence of carbon dioxide in the Earth's atmosphere can be explained by the Earth's oceans. Venus is too close to the Sun for water to exist as a liquid. At Mars' distance from the Sun, water is locked up as ice. But for the Earth, as for Baby Bear, everything "was just right." As soon as the Earth's crust cooled sufficiently, the basins filled up with water.

Carbon dioxide readily combines with water to form carbonic acid. Carbonic acid in solution in the Earth's oceans loses its hydrogen and combines with calcium and magnesium to form insoluble carbonates. The carbonates precipitate out of solution into bottom sediments which harden into the rocks we know as limestone and dolomite. Animals and plants in the sea take up carbonates to build their skeletons or shells. At death, these hard parts sink to the bottom of the sea to become part of the carbonate sediments. The carbon dioxide in the Earth's early atmosphere and most of the carbon dioxide and carbon monoxide that has been expelled volcanically from the Earth's interior is locked up in the rocks of the Earth's crust. The limestone quarries of Indiana or England's chalk cliffs of Dover are the sort of places to look for the Earth's early carbon dioxide atmosphere.

And what about the oxygen, where did the oxygen in the Earth's atmosphere come from? Some of the oxygen resulted from the breakup of water molecules in the upper atmosphere by ultraviolet light (the hydrogen subsequently escaped to space). But most of the oxygen in our atmosphere was contributed by life, as a by-

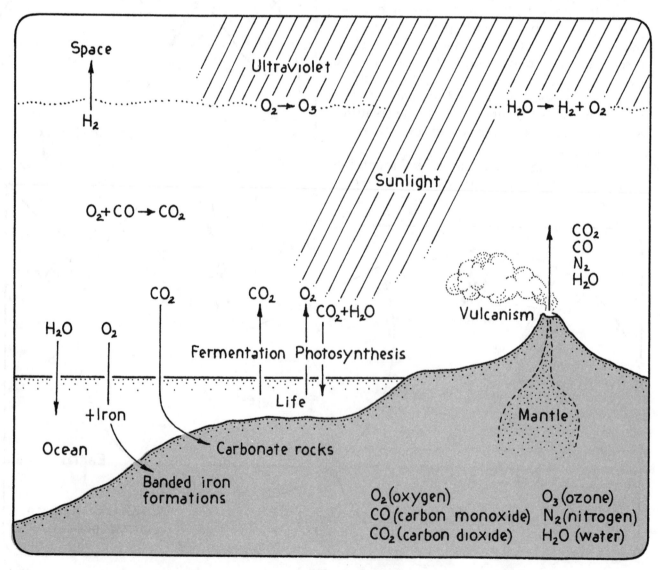

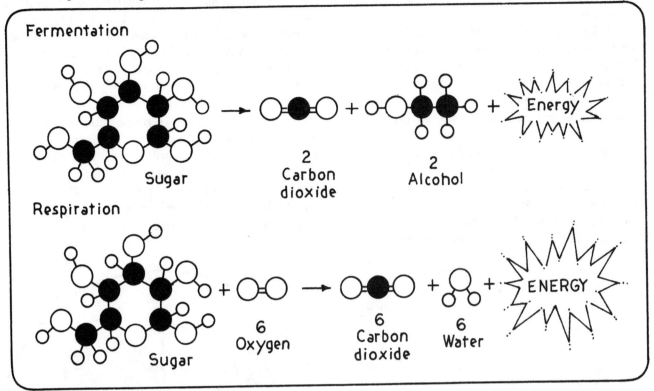

Fermentation

Sugar → 2 Carbon dioxide + 2 Alcohol + Energy

Respiration

Sugar + 6 Oxygen → 6 Carbon dioxide + 6 Water + ENERGY

product of photosynthesis. The blue-green bacteria, busy packing sunlight into sugar, manufactured the air we breathe.

In the sketch at left, I have tried to summarize the major processes that helped transform the Earth's early atmosphere. The primeval atmosphere came out of the ground volcanically, and was probably pretty much the same on Earth as on Mars or Venus. But water and life have changed the Earth's primitive assignment of gases beyond recognition. Only the relatively inert gas nitrogen was immune to the machinery of transformation. In time, nitrogen became the dominant component of the Earth's airy envelope.

The banded iron formations disappeared from the geological record about 2 billion years ago. At that time, the iron in the Earth's seas had been fully oxidized, and free oxygen in the atmosphere began to rise unchecked. Estimates of the amount of oxygen in the Earth's atmosphere 2 billion years ago range from 1 percent to 50 percent of the present level. Whatever the percentage, the gas posed a deadly threat to living organisms.

As oxygen levels rose, many photosynthesizing bacteria may have become extinct, burned up by the pernicious byproduct of their own light-trapping success. Other bacteria retreated to muddy lake bottoms or the lower layers of mat communities to escape the toxic gas. Oxygen was the "Catch-22" of photosynthesis.

But as one might expect, after more than a billion years of the struggle to survive with ever increasing levels of oxygen, life evolved defenses. Organisms developed special enzymes to react with the dangerous molecular units produced by oxygen and convert them into harmless carbon compounds and water. But enzymes were only a stopgap defense. A permanent solution to the oxygen crisis was respiration.

Respiration is a kind of controlled burning. Instead of letting oxygen "burn down the house," the cell evolved a kind of "fireplace" where oxygen could be used to burn fuel safely. Once this potentially dangerous technique had been sorted out chemically inside the cell, life was safe from oxygen. At the same time, life discovered a dazzlingly efficient way to get energy from sugar.

For a billion and a half years life had been producing sugars by photosynthesis and releasing the energy stored in sugars by fermentation. Fermentation is not a very efficient way to get the energy out of sugar. Much of the energy remains locked up in the alcohol molecule which (for everyone but the brewer) is a waste product of the process.

Worse, alcohol is a poison, as some of us have realized to our sorrow the morning after. The early fermenting bacteria were not equipped to deal with too many "morning afters." But as long as the bacteria stayed in the sea, water could dilute and wash away the alcoholic waste of fermentation.

Respiration solved all the problems of fermentation.

By channeling oxygen in the environment to the burning of sugar, an enormous advantage was gained. First, the oxidation of sugar breaks that molecule down into smaller units and re-leases much more of the stored energy. Fermentation releases about 110 calories from a gram of sugar. Respiration releases almost 3900 calories from the same gram. From a single unit of food, respiration provides 35 times more useful energy than fermentation. And to make a good thing better, the waste products of respiration are innocuous carbon dioxide and water.

The development of respiration was a turning point for life. Life entered the fast track. The combination of photosynthesis and fermentation was a kind of "subsistence farming," producing barely enough energy for survival. Respiration produced a surplus of energy, a capital that could be put to creative advantage. Now the

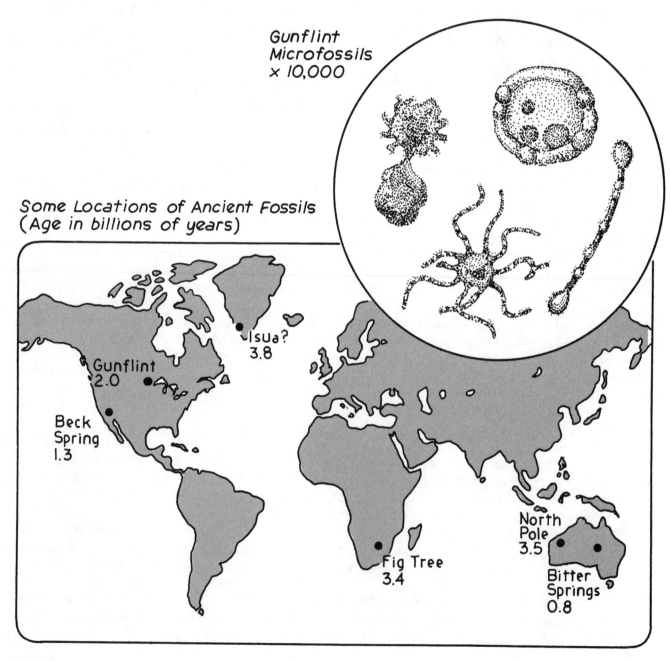

Gunflint
Microfossils
× 10,000

Some Locations of Ancient Fossils
(Age in billions of years)

Isua? 3.8

Gunflint 2.0

Beck Spring 1.3

Fig Tree 3.4

North Pole 3.5

Bitter Springs 0.8

barn was full and life could shrug off an occasional dry season or hard winter.

Bacteria that get their energy by fermentation still exist, but they never amounted to much. They never adapted to oxygen and they must shun fresh air. They cling to a tentative existence in hidden places—stagnant water, sewage treatment plants, the intestinal tracks of animals, ocean bottoms, and hot springs—any place they can escape the deadly oxygen. Fermentation was an evolutionary dead end.

Two billion years ago the air breathers moved to center stage. Respiring cells were the new wave of evolution. Soon they dominated every available habitat. In oceans, lakes, and streams, hot water and cold, the sunlit surfaces of ponds and the muddy bottoms, the respiring bacteria were everywhere. With a sunny day, a few basic chemicals, and carbon dioxide, life now made its own fuel by photosynthesis. With a little oxygen, the fuel was burned completely to yield a surplus of energy. Oxygen had become an asset rather than a liability.

An impressive variety of microfossils appear in 2-billion-year-old rocks. The best known fossils from that time are in the Gunflint cherts of the Lake Superior region of North America. Cherts are tough, fine-grained sedimentary rocks composed of silica. The mineral precipitates quickly from seawater and hardens with equal rapidity. It becomes a durable glassy casket for whatever cells are trapped in the sediments.

The Gunflint cherts contain an inventive array of cellular forms: flowers, stars, filaments, rods, clustered spheres, umbrellas, and parachutes. The fossil record shows that life used its new energy surplus for a wave of experimentation.

New life cycles emerged. Some bacteria budded off progeny cells that swam away to find new homes before they came to resemble their parent. Others formed webs of filaments that twined through soil. Still others formed branches that scattered spores to the winds. The most important of all the innovations was a new strategy for reproduction, a strategy involving two parents. The invention of sex changed life on Earth forever.

A third of Earth's history had passed between the invention of photosynthesis and the invention of respiration. Obviously, some things happened during those "dark ages." The Earth's crust was evolving. Continents were growing. The oceans were becoming saltier. Cyanobacteria were developing new forms, new means of mobility, new defenses against oxygen, new repair mechanisms to cope with the damage caused by ultraviolet radiation.

But progress was slow. Reproduction was by simple cell division. Each pair of offspring were exact copies of the parent, and carried an exactly congruent complement of DNA. Only the occasional random mutation of the DNA code sequence led to novelty, a mutation due to cosmic ray damage or the disturbing effects of natural radioactivity or ultraviolet light. Most mutations did nothing to enhance the survival ability of the cell, and the change died without a trace. The rare mutations that gave a cell a competitive edge tended to endure and copy themselves into the growing diversity of life.

Sometime after the invention of respiration, life used part of its newly won capital to create a new kind of cell and a new method of reproduction. With these events, to be recounted in the next chapter, the pace of evolution accelerated a hundredfold.

Powell Plateau

Paleozoic Strata

Grand

Canyon

Series

Vishnu

Schist

Colorado River

Enter, the Supercell

The naturalist Joseph Wood Krutch said the Grand Canyon was "the most revealing single page of Earth's history anywhere open on the face of the globe." It would be hard to disagree. I have never felt the experience of delving so deeply into the literature of time than on a visit to the Grand Canyon. Here geological circumstance has laid bare two billion years of Earth history.

Northern Arizona was once a region of low elevation where the ancestor of the Colorado River meandered sluggishly to the sea. Then, tens of millions of years ago, this part of the Earth's crust began to rise, pushed straight up from below like the floor of an elevator. As the crust went up, the river cut down into the rock, struggling to maintain its ancient course. The uplift and the cutting continues today. The incision has now reached well over a mile into the crust, exposing ever more ancient layers of sedimentary rock, each with its own story to tell of life on Earth. At the bottom of the gorge the river has sliced into the bones of the continent itself, into rocks half as old as the planet.

The oldest formations exposed in the Grand Canyon are the Vishnu schists, metamorphic rocks formed by great heat and pressure from even more ancient sedimentary and volcanic formations. They are severely deformed, crisscrossed by faults, and squeezed into folds. They contain no recognizable fossils.

But the Vishnu schists do confirm that two billion years ago the crust of the Earth was active. Sea levels rose and fell relative to the land. Lavas gushed from the Earth's interior. Continents heaved together and pushed up mountains. Weathering and erosion pulled the mountains down and piled the gritty residues in sedimentary basins. There is evidence of massive glaciations during this same era. Amidst all this fire and ice and unending violence, single-celled bacteria struggled to cope with an unstable environment. The pressure toward adaptive change was overwhelming.

The Vishnu schists are overlaid by tilted sedimentary strata known as the Grand Canyon Series. These rocks are about a billion years old. A long chapter of uplift and erosion separate their deposition from the creation of the Vishnu schists. The Grand Canyon Series continue the story of intense crustal activity. The strata contain few evidences of life. This is not because life did not exist, but because soft-bodied single-celled organisms do not readily fossilize. There were no bony creatures in those days, no organisms with shells or backbones. There was

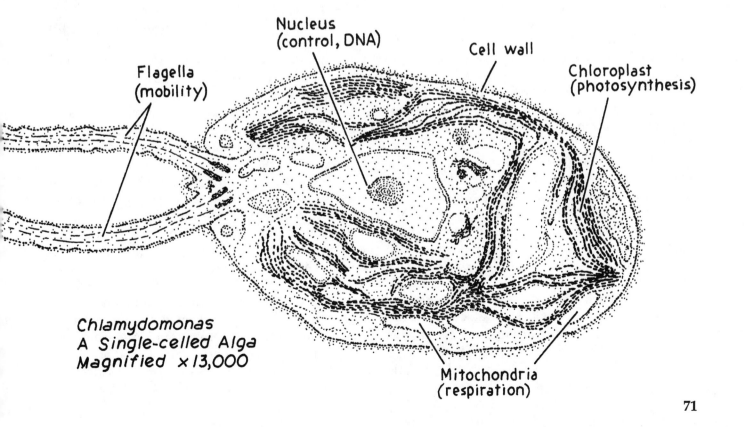

Flagella
(mobility)

Nucleus
(control, DNA)

Cell wall

Chloroplast
(photosynthesis)

Mitochondria
(respiration)

Chlamydomonas
A Single-celled Alga
Magnified x 13,000

only a film of photosynthesizing bacteria reaching for the sun, and a scum of oxygen-shunning fermenting bacteria cowering in nooks and crannies.

As long as evolution depended upon random mutations of the genes to effect change, organisms were at the mercy of geological violence. In many cases, environments changed more rapidly than organisms could adapt.

But change did occur. The era of the Grand Canyon Series saw the development of a new kind of cell, a compartmentalized supercell with special organs for special functions, a resourceful cell that combined the best features of its predecessors. I have represented the new kind of cell with a modern alga, the *Chlamydomonas*. The new cells are called eukaryotes, meaning "having a true nucleus." The spectacularly successful innovation swept all else before it. All multicelled creatures alive today are eukaryotic.

The key feature of the new cell was the nucleus. The nucleus contains the precious DNA material safely packaged within a membrane. The nucleus is the control room of the cell, where the genetic blueprints are stored and copied. In the older-type cell (called prokaryotes, "before a nucleus"), the genetic material lay naked and diffused throughout the body of the cell.

It is uncertain why the genetic apparatus came to be encapsulated within a membrane. It may be that the membrane helped protect the sensitive genetic codes from the destructive effects of oxygen. Certainly, the new more complicated cell had to carry a fuller package of blueprints. Perhaps the enveloping membrane helped keep the longer strands of DNA untangled during the replication and distribution of genetic information.

The success of the membrane in helping the DNA do an untangled dance of replication allowed for a huge increase in the amount of information the genes could carry. The DNA strand in a human cell, if stretched out, would be a yard long. By contrast, the DNA strand in a prokaryotic bacterium is no longer than this letter *i*.

I have labeled several other distinctive features of the new-type cell. The mitochondria are the power packs, the furnaces where fuel is burned, the centers of respiration. There may be hundreds of mitochondria in a typical cell. The

chloroplasts are the light-trappers, the workshops of the green chlorophyll, the centers of photosynthesis. In the alga *Chlamydomonas*, there is one large cup-shaped chloroplast enclosing the nucleus. A green plant cell might contain as many as 50 chloroplasts. Finally, there are the flagella, the whiplike appendages that give the cell mobility, the outboard motors, the tadpole tails (see Note). *Chlamydomonas* has two flagella. Each of the components of the eukaryotic cell has its own tool kit of special enzymes, adapted to exact needs.

The division between the simple prokaryotes and the multi-compartmented eukaryotes is the greatest gulf among living organisms, even more basic than the distinction between plants and animals. The new cells were a giant leap forward for life, and there was no looking back. Although the prokaryotes dominated the Earth for the first 2 billion years of the history of life, they are represented today only by bacteria.

Eukaryotic cells seem to have appeared quite suddenly—geologically speaking—about 1.5 billion years ago. No transitional forms have been found in the fossil record. It is as if the Wright brothers' Kitty Hawk flying machine was followed a week later by the Concorde jet. The abruptness of the transition and the amazing increase in the complexity of the new cells has long baffled biologists.

Biologist Lynn Margulis has revived an old idea to account for the origin of the eukaryotes. Her theory is a story of hard times and cooperation. According to Margulis, the eukaryotic cells arose by symbiotic combinations of single-celled prokaryotes. Symbiosis is a partnership of organisms for mutual benefit. The gist of the theory is outlined with the diagram on the following page.

The first step in the genesis of the new cell was the inclusion of an oxygen-respiring bacteria into the body of a fermenting host cell. Initially the relationship may have been that of predator to prey, the fermenter ingesting the respirer. Then the relationship became that of host to guest, when the fermenting host cell decided to exploit the talents of the respirer rather than digesting it. The host cell may have evolved a nuclear membrane to protect its DNA from the oxygen required by its houseguest. The host was then free to use the excess energy provided by

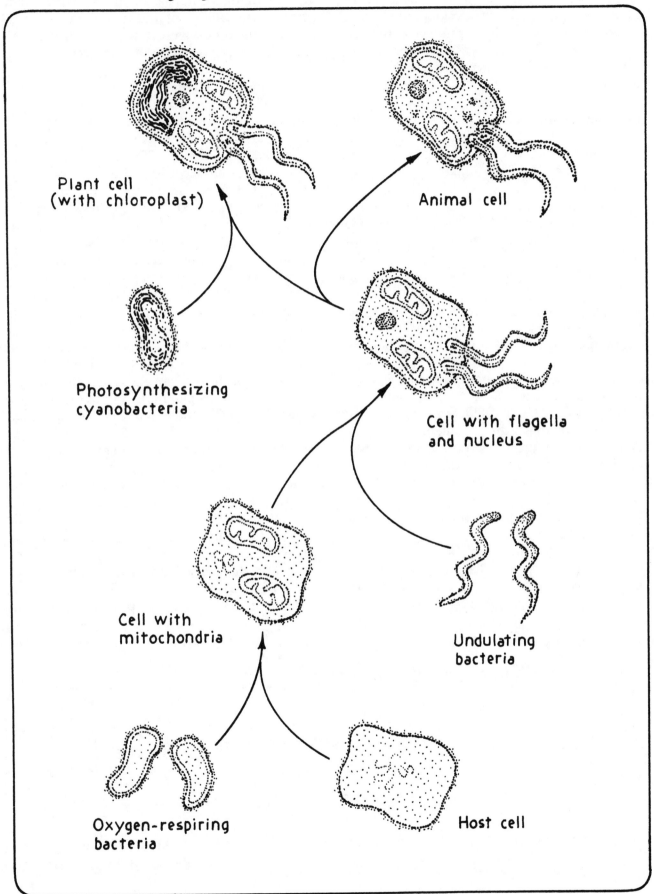

Plant cell
(with chloroplast)

Animal cell

Photosynthesizing
cyanobacteria

Cell with flagella
and nucleus

Cell with
mitochondria

Undulating
bacteria

Oxygen-respiring
bacteria

Host cell

the guest's efficient respiring apparatus. The small oxygen-hungry guest had the advantage of access to the food-rich interior of the host cell, a pantry crammed with the produce of the host's fermentation. The guest was protected from the rigors of a life on its own, and the host could take indirect advantage of the growing oxygen content of the atmosphere. Cooperation served both parties. Eventually the guest cell was fully incorporated into the host as a mitochondrion.

The next step, according to the symbiotic theory, came when threadlike undulating bacteria called spirochetes, the Olympic swimmers of the bacterial world, attached themselves to the evolving "supercell." The swimmers were after food leaking through the host cell's outer wall. But they were not unwelcome parasites. Like swimmers hanging on the back of a raft and kicking their feet, the undulating spirochetes gave the host cell locomotion. With full incorporation, the hangers-on became flagella. (This account of the origin of flagella is highly controversial.)

A third symbiotic connection came with the invasion of a host cell by a photosynthesizing cyanobacteria. The ingested sun-worshiper gave the host cell the indirect benefits of sunlight chemistry. The guest derived the same benefits as the ancestors of the mitochondria. Again, what began as cooperation for mutual benefit evolved to become an essential dependence. Guest became chloroplast and the first true plant cells appeared on Earth.

The mitochondria and chloroplasts in modern cells still carry small strands of their own DNA. They still clutch to themselves those emblems of their former independence.

The rapidity with which these develop-ments took place may perhaps be explained by the changing composition of the atmosphere and the instability of the geological environment. The times were hard. For prokaryotic cells, it made sense to pool resources.

Symbiotic adaptations were more readily accomplished than those which required changing the genes. As an ozone layer built up in the upper atmosphere, ultraviolet radiation diminished at the surface. The rate of genetic mutations dropped accordingly. New ways of ensuring adaptive variation were required, and the new supercells quickly evolved a new method of sexual reproduction. Sex insured variation among offspring, not on a random basis, but systematically, generation by generation.

Prokaryotic cells reproduce by binary fission. Each cell has a single circular strand of DNA containing the blueprint for the next generation. When the time comes for reproduction, the DNA copies itself, the mirrored strands separate, and a wall extends to divide the cell into two identical offspring. The offspring are exact copies of the parent, except in the rare case when the DNA of the parent cell has undergone a mutation. Populations that reproduce by binary fission are inherently stable.

The new eukaryotic cells perfected the fission process into a marvelous reproductive ballet. This may have been necessary to handle the replication of the much longer DNA blueprint required by the supercells. Most eukaryotes carry about a thousand times more genetic information than their bacterial ancestors.

The DNA molecules in eukaryotic cells are packed up with RNA and proteins into neatly paired bundles called chromosomes. All

Cell Reproduction by Binary Fission

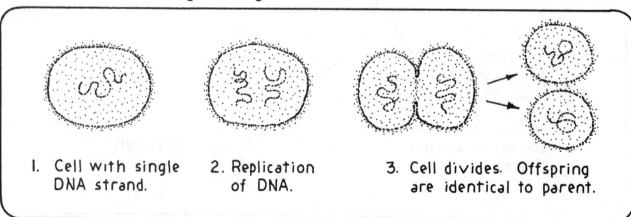

1. Cell with single DNA strand.

2. Replication of DNA.

3. Cell divides. Offspring are identical to parent.

Cell Reproduction by Mitosis

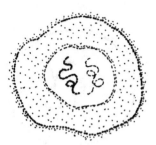

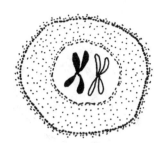

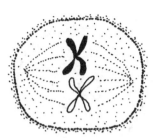

1. Chromosomes are extended threads in nucleus.

2. Chromosomes replicate and coil up. The copies are joined at some point along their length.

3. Nuclear membrane breaks down, chromosomes line up independently.

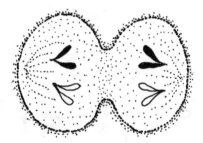

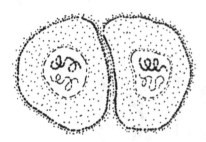

4. Doubled chromosomes break and pull apart.

5. Nuclear membrane forms, cell divides, chromosomes uncoil. Offspring are identical to parent.

eukaryotic cells have at least two chromosomes. Human cells have 23 pairs of chromosomes. The process by which the cell divides to make copies of itself is called mitosis. In my diagram of mitosis I have shown the *pas de deux* of only one pair of chromosomes. In the human cell, 23 chromosomal couples line up to form the *corps de ballet*.

This elegant reproductive dance is assisted by an elaborate molecular machinery. Margulis believes that the system of microscopic tubes (called the spindle) that pull the paired chromosomes toward opposite poles of the cell (steps 3 and 4) are a remnant of filamentary bacteria that were incorporated into the machinery of the cell by the same sort of symbiotic relationship that gave rise to the flagella. She also believes that the double complement of chromosomes in eukaryotic cells may have arisen from acts of cannibalism during hard times.

Mitosis is the process by which some single-celled eukaryotes reproduce. And it is the way multicelled organisms grow and generate special cells for special purposes—for example, tissue, bone, blood. It was a variation on mitosis, called meiosis, that introduced sex into the history of life on Earth.

In meiosis, the paired chromosomes in the parent cell are distributed among the offspring, possibly with some mixing of genes along the way. The end result is four offspring, each with one-half the number of chromosomes of the parent. In humans, for example, the offspring of meiosis are the egg and sperm cells—depending on the sex of the parent—with 23 chromosomes apiece. A sex cell from one individual unites with a sex cell from a parent of the opposite sex to restore the original number of chromosomes and create a new individual.

This complex choreography of reproduction

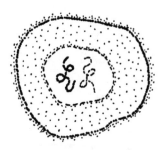 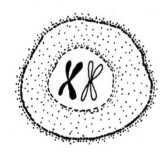 

1. Chromosomes are extended threads in nucleus.

2. Chromosomes replicate and coil up. The copies are joined at some point along their length.

3. Nuclear membrane breaks down, chromosomes line up by pairs. Possible exchange of genes.

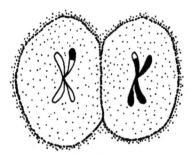 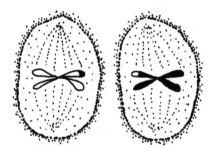

4. Chromosomes pull apart but do not break. Cell divides.

5. Chromosomes line up as in mitosis and pull apart.

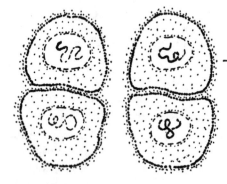 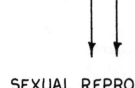

6. Nuclear membrane forms, cells divide, chromosomes uncoil. Each of four sex cells has half the parent's number of chromosomes.

7. SEXUAL REPRODUCTION Sex cell from one parent unites with sex cell from other parent to form new cell with full number of chromosomes but unlike either parent.

had but one justification—systematic variation. Offspring of sexual reproduction receive different combinations of the parental chromosomes. Advantageous traits from different individuals are brought together quickly. The pool of genes is stirred every generation. Beneficial variations are selectively reinforced by the struggle to survive on a precarious planet.

Half a billion years or more may have been required for these elaborate reproductive mechanisms to become established in the new eukaryotic cells. But once sex was implanted in the stream of life, there was no holding back the drive toward diversification. To use a human metaphor, a traditional society dedicated to preserving the old ways yielded to a society based on systematic innovation.

Our climb out of the Grand Canyon brings us at last to the horizontal Paleozoic strata, laid down beginning about 600 million years ago. Suddenly the fossil record becomes rich and various. Innovation abounds. The sexually-reproducing eukaryotes have inherited the Earth.

The Great Acceleration

About 600 million years ago, multi-celled life forms appeared with surprising suddenness. These included the first creatures with hard shells. The Earth's greatest ice age may have spurred the wave of innovation.

In the winter of 1850 a fierce storm smashed into the Orkney Islands off the northern coast of Scotland. Roaring wind and towering waves lashed the beaches and stripped away sand hills along the shore. In their fury the elements unearthed a stone village that had lain buried beneath the dunes for four thousand years. The place is called Skara Brae.

The village at Skara Brae is not unique. There are many similar ruins along the western shores of Europe. What is special about Skara Brae is its remarkable state of preservation. The sands drifted over the prehistoric settlement shortly after it was abandoned. Almost no stone has been toppled from its original placement. Only the roofs of the huts—skins, perhaps, over whalebone rafters—are gone.

The plan of Skara Brae is shown below. The village contains about ten stone huts, roughly circular, connected by outer walls and inside passages. Hearths, beds, benches, and pantries are all of stone. The clustered arrangement was, no doubt, a snug refuge against the elements and a firm redoubt against marauders.

The huts at Skara Brae show similarities to earlier stone structures at Malta in the Mediterranean and along the coasts of Spain. It is not unlike the medieval stone monastery we visited on Skellig Michael in an earlier chapter. Once this pattern of building became established on Europe's western seaboard, it remained fixed for a hundred generations. It was well suited for life on the wild Atlantic shore.

It is possible to think of Skara Brae as a kind of fossil, a stony shell exhumed from the earth. The fossil shell once enclosed a living organism. The organism was made up of dozens of human individuals sheltering within circular cells. The cells were linked by passages for communication and the transfer of food and fuel. The cluster was enclosed by a protective outer membrane.

On the remote, storm-swept Orkney Islands humans grouped their cells together of sheer necessity. An assembly of cross-linked cells was well suited to withstand the rigors of the harsh Atlantic environment. We sometimes applaud this kind of cooperation as civilized behavior. In the Orkneys, cooperation was a strategy for survival.

In *The Lives of a Cell* Lewis Thomas wrote: "If it is in the nature of living things to pool resources, to fuse when possible, we would have a new way of accounting for the progressive enrichment and complexity of form in living

Plan of
Prehistoric Village
at Skara Brae

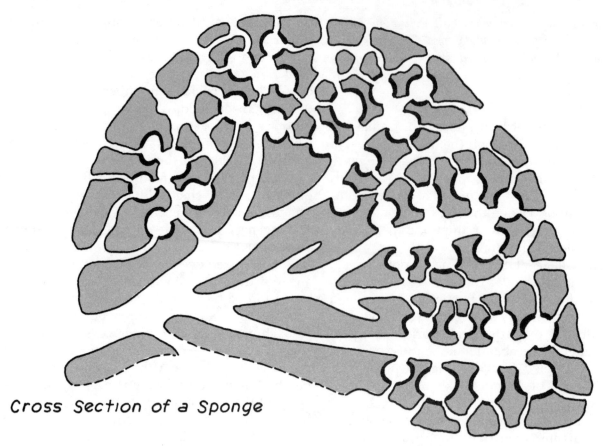

Cross Section of a Sponge

things." Thomas muses that the tendency to cluster for mutual benefit may have worked on several levels of evolution.

One must be careful about the flippant use of metaphors, or of ascribing willful action to molecules or microbes. And yet, I cannot think for long about the evolution of life on Earth without supposing that even atoms are somehow charged with a passion to become an eagle or a geranium. Atoms unite to build proteins and sugars and fats and nucleic acids. Large organic molecules are the building blocks of prokaryotic cells. Prokaryotes throw in their lot together and become eukaryotes. Eukaryotes pool specialized talents to become sponges and flatworms and slime molds and seaweeds. And so the dance of life whirls toward greater and greater complexity, against all reasonable probability and the counter-urge of entropy.

I am struck by the similarity of the cross section of a sponge to the plan of Skara Brae. I like to imagine that the two have something in common. Sponges are among the earliest forms of multicellular life. They are not true multicellular animals, with diverse specialized cells. Rather, they consist of a single type of cell with the chameleon ability to change its form to suit its task. Inward-facing cells are equipped with

whiplike appendages. The concerted beating of these tiny threads pumps water through the sponge. Other cells act as the building blocks of the sponge's outer wall. Still others—the "rubble between the walls"—provide some measure of support. Acting together, the cells of the sponge build an efficient mechanism for sieving nutrients from water.

Like the village at Skara Brae, the sponge was an early attempt at pooled resources, a first try at communal living. Sponges have evolved little since they appeared in the seas 600 million years ago. They represent—in an archeological sort of way—the inevitable urge of life toward multicellularity. The higher animals are to the sponge as the great cities of the world are to Skara Brae.

Multicelled organisms appear in the geological record with surprising abruptness. Within a few hundred million years—a few ticks of geological time—single-celled microscopic ancestors gave rise to a zoolike range of large animals. Seven hundred million years ago a visitor from another planet would have required a microscope to examine life on Earth. Five hundred million years ago the seas swarmed with sponges, trilobites, molluscs, brachiopods, and worms, creatures that crawled on legs or sailed like sloops through algae seaweed forests of red and green and brown. The sudden appearance of these creatures marks a bold discontinuity in the fossil record of life.

As geologists of the 19th century began to read Earth history from the record of the rocks, they assigned ages to the strata on the basis of the fossils they contained. The names assigned to periods of the geological past—Devonian, Jurassic, Pennsylvanian, for example—often derived from the place where the representative formations were first studied: Devon in England, the Jura Mountains in Germany, Pennsylvania.

The oldest strata that contained recognizable fossils were termed *Cambrian*, after the Roman name for Wales. Still older rocks showed no signs of life, and geologists could only suppose that life must have appeared on Earth shortly before the onset of the Cambrian age. Everything that preceded that moment in the geological record was termed *Precambrian*.

As time passed the true age of the Earth was recognized, and the Precambrian era grew to encompass most of Earth history (see our familiar time-line mollusc, page 82). But the idea that life appeared on Earth only toward the end of the Precambrian era was an illusion that persisted into our own century. Within recent decades, however, the subtle evidence for microscopic life has been found in very ancient rocks, rocks much older than the Cambrian. The origin of life on Earth has been pushed back billions of years.

In the 1950s the geologist Martin Glaessner discovered a sequence of strata in the Ediacara Hills of South Australia that preserved a continuous record of life across the Precambrian-Cambrian divide. The circumstances of deposition of the Ediacarian strata allowed for fine impressions of soft-bodied creatures with forms like worms and jellyfish. The Ediacarian fossils represent early experiments in multi-cellular living. Similar fossils have subsequently been found worldwide.

It is now clear that the Precambrian-Cambrian boundary does not mark the origin of life, but rather an acceleration of evolution toward large multi-celled forms. During the late Precambrian and early Cambrian times, multi-celled life forms diversified into virtually all the present families of life on Earth.

What caused this sudden explosion of complexity and diversity? There is no clear answer to the riddle, but there have been many educated guesses. The guesses all involve a close interplay of biology and geology.

Biologically the stage was set for the great acceleration with the development of sexually reproducing eukaryotic cells. Sex meant diversity, a mixing of the genes, a roll of the dice with every generation. The sophisticated internal organization of the new "supercells" allowed for the maintenance of a larger genetic blueprint, strands of DNA as long as your arm.

The selective pressure toward rapid evolution was supplied by the geological instability of the times. Previously unoccupied ecological niches were opened up to small populations of novel organisms. The "times were a-changin'," as the saying goes, and innovation was encouraged. Increasing levels of oxygen in the atmosphere favored experiments in cooperative living. Ozone screened more of the Sun's ultraviolet radiation, and life near the surface of the sea and in shallow waters became feasible.

The geological record hints at another dramatic modification of the physical environment. On the eve of the great acceleration, there occurred a worldwide catastrophe that placed enormous pressure on all life, closing some ecological niches and opening new ones. The catastrophe was the greatest ice age of all time.

We are in the midst of an ice age today, although in something of an intermission (see Chapter 20). The past few millions of years have been abnormally cold. On numerous occasions during that time the ice has crept down from arctic centers of accumulation to burden the northern continents. Only 15 thousand years ago Boston, New York, and Chicago were under the ice. Greenland and Antarctica are still locked in the deep freeze.

Farther back, there are hints in the geological record of at least four other eras characterized by massive ice sheets. The most extensive of these seems to have occurred between 900 million and 600 million years ago, with at least three major pulses of glaciation. This late great Precambrian ice age may have been part of the push that sent life tumbling toward diversity.

Geologists reconstruct ancient ice ages on the basis of several kinds of evidence. One common clue is scratches and grooves on the surfaces of rocks that were exposed at the time of

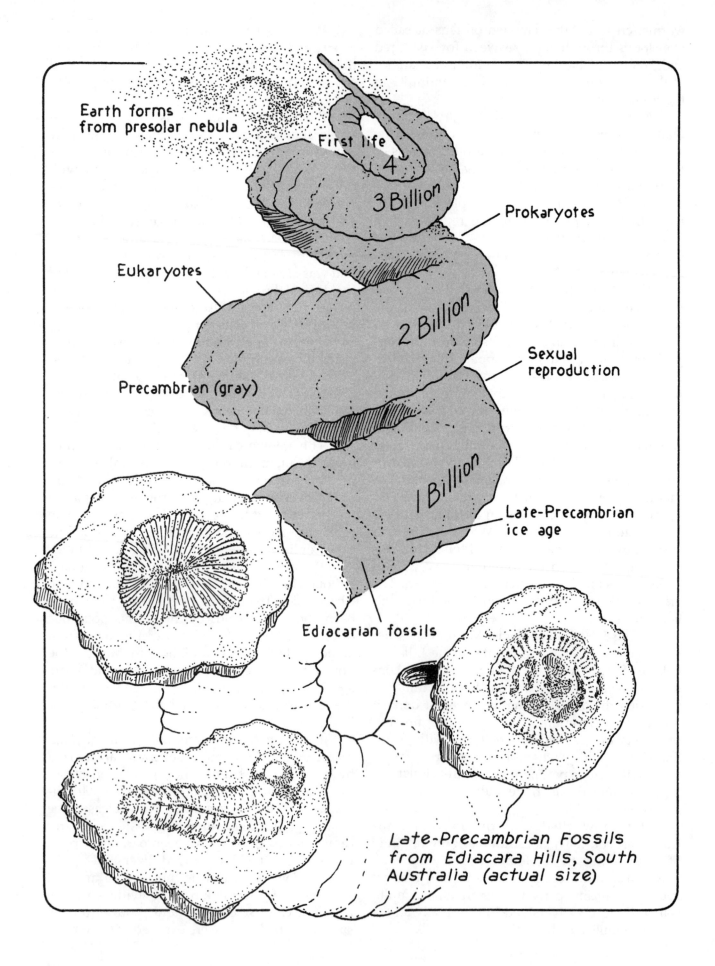

Earth forms from presolar nebula

First life

4

3 Billion

Prokaryotes

Eukaryotes

2 Billion

Precambrian (gray)

Sexual reproduction

1 Billion

Late-Precambrian ice age

Ediacarian fossils

Late-Precambrian Fossils from Ediacara Hills, South Australia (actual size)

the glaciations. The scratches speak eloquently of mile-thick sheets of moving grit-filled ice. A second clue is extensive formations of a rock called tillite. Tillite is a sedimentary formation derived from the erosional debris deposited by a retreating glacier. It consists of sand, clay, pebbles, and boulders scoured from the landscape by moving ice, carried perhaps great distances, rounded and smoothed by abrasion, and dropped where the ice melted. Tillite is glacial rubble which time, burial, and chemical action have consolidated into solid rock.

Evidence for late Precambrian glaciations has been discovered on every continent. The ice sheets which blanketed the Earth in those times seem to have extended even into tropical latitudes. The late Precambrian Jack Frost touched the entire planet with his icy brush.

The cause of the late Precambrian ice age is not known. Geological turmoil of the Earth's crust could have initiated glaciation by changing the composition of the atmosphere or altering the course of ocean currents. Or perhaps the Sun faltered. Perhaps the Solar System passed through a dusty region of the galaxy and sun-light was blocked from reaching Earth. Perhaps a decrease in atmospheric carbon dioxide "shattered the glass" that had maintained a mild greenhouse effect and allowed more of the Earth's heat to escape to space. The Earth's climate is as finely tuned as a Swiss watch. The continents can be tipped toward an age of ice by small adjustment of any one of a dozen gears or levers.

It is by no means certain that the late Precambrian glaciations were the cause of the great biological acceleration. But there can be little doubt about the selective pressure of a great ice age. As ice built up on the continents, sea levels dropped. Continental shelves were left high and dry. When the ice melted, sea levels rose again. Sea temperatures rose and fell with each ebb and flow of the ice. Ecological niches were created and destroyed. This instability, coupled with rising levels of oxygen and the development of sexually reproducing eukaryotic cells, set off an explosion of biological innovation.

The first experiments in multicelled living were soft-bodied creatures represented by the Ediacarian fossils from Australia, wormlike and jellyfishlike creatures that floated in Ediacarian

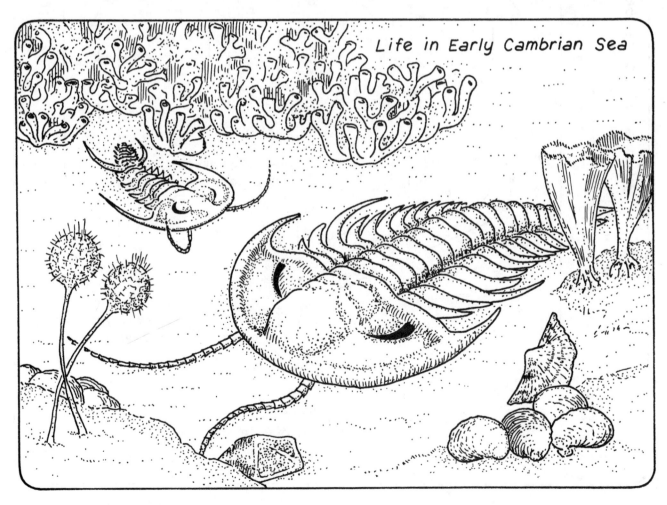

Life in Early Cambrian Sea

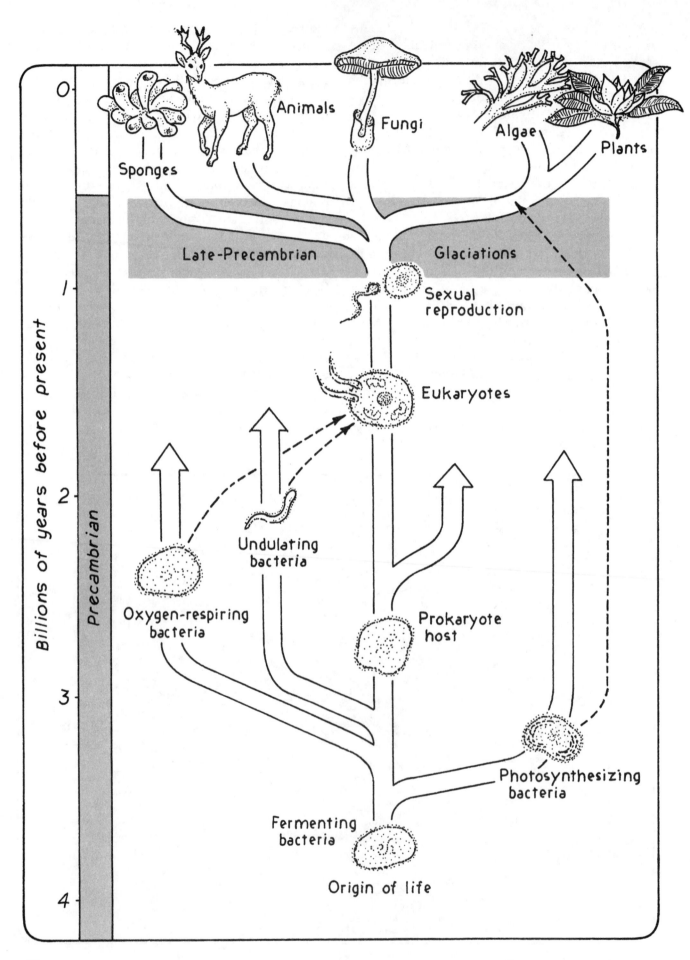

Billions of years before present

Precambrian

0 —
Sponges
Animals
Fungi
Algae
Plants

Late-Precambrian   Glaciations

1 —
Sexual reproduction

Eukaryotes

2 —
Undulating bacteria

Oxygen-respiring bacteria

Prokaryote host

3 —
Photosynthesizing bacteria

Fermenting bacteria

Origin of life

4 —

ponds and seas or burrowed on muddy bottoms. These delicate pioneers established once and for all the benefits of pooled resources.

The most successful innovation, aside from multicellularity, was the development of external shells for protection and support. As oxygen levels increased and circulatory systems evolved, the surfaces of cells no longer had to be in direct contact with the environment for the cell to procure oxygen and nutrients. Creatures could gird themselves with plate. Shells had obvious survival value. They were armor against assault. They provided points of attachment for muscles which made possible new modes of articulation and movement.

Shells fossilize easily. And so it was that rocks of the Cambrian era, 570 million years old, quite suddenly seemed—to 19th-century geologists—to be crammed with fossils. There were molluscs and brachiopods, seashells like the ones you pick up on the beach today. There were shrimplike arthropods with crusty husks.

But most spectacular of all were the trilobites.

The trilobites take their name from their armored bodies which were divided into three segmented lobes. I have often picked up fossil trilobites an inch or so long. The giants of the family were as long as a foot. These were the battle cruisers of the shallow Cambrian seas, the bottom-scavengers and mud-eaters, the lords of their environment. I have illustrated two early trilobites on page 83. These splendid creatures, like the dinosaurs or humans, might appropriately have given their name to their era.

By the end of the great acceleration, all of the present families of life were in existence. Plants, fungi, animals, and sponges had found their niches and perfected their special ways of making a living. The climate moderated, the ice fell into retreat, the seas warmed and teemed with visible life.

But the land was still bare. The surfaces of the continents were moonscapes. Formidable barriers remained to be conquered before life could move ashore.

The Invasion
of the Land

The Earth's canopy of air protects life against dangers from space. Once an ozone layer was in place it became feasible for life to move out of the water onto the land.

Few pleasures are as sweet as the pleasure of lying on a grassy slope beneath a dark August sky and experiencing a night full of shooting stars.

The greatest of the annual meteor showers climaxes near the 12th of August. The shower is called the Perseids, and if you watch carefully you will see that the "shooting stars" seem to radiate from the constellation Perseus.

My drawing is a view of the northeastern sky late in the evening of August 12 from the latitude of Ireland. The Milky Way plunges down across the night, from the W-shaped constellation of Cassiopeia to the brilliant star Capella in the constellation Auriga. Between Cassiopeia and Auriga, astride the Milky Way, are the stars of Perseus, hero of Greek myth, slayer of the snaky-haired Medusa, rider of the winged horse Pegasus, rescuer and lover of the beautiful Andromeda. The brightest star of Perseus is Mirfak, and near this star is the radiant center of the Perseid shower.

I have drawn several meteors streaking across the sky. Of course, it is highly unlikely that more than one meteor would be visible at

the same time. But if you keep your eyes open you may see several dozen in an hour.

These "shooting stars" or "falling stars" are not stars at all, but tiny fragments of a disintegrated comet. They are typically no larger than a grain of sand. The particles move in a parallel stream along the old orbit of the comet, and in August the Earth cuts across the stream. As the particles collide with the Earth's atmosphere they are heated by friction and are vaporized. It is the streak of glowing vapor that we see as a "shooting star."

The Perseids are a particularly reliable shower of "shooting stars," but you can see meteors on any clear night. It has been estimated that hundreds of millions of visible meteors enter the Earth's atmosphere each day and perhaps billions more that are not visible. All but the largest of these are vaporized. If it weren't for the atmosphere, the surface of the Earth would be subjected to a slow but steady rain of sand from the sky.

Even if unchecked, this rain of stone is too sparse to be a serious threat to life. Humans have survived quite well on the moon without a canopy of air. But the work of the meteors is readily apparent on the moon. I have reproduced two famous Apollo photographs of lunar footprints. It is clear that the moon's surface is

covered with a powdery dust. Where did the dust come from if there is no wind or water to erode the moon's solid rock? The dust is the result of four billion years of bombardment of the lunar surface by meteorites of all sizes. The entire surface of the moon has been pulverized by "shooting stars."

Although a gentle rain of meteorites might not threaten the survival of any species of life on Earth, an individual animal would not wish to be struck by a grain of sand traveling at a hundred thousand miles per hour. I, for one, am grateful for our umbrella of air.

The atmospheric umbrella is surprisingly thin, compared to the size of the planet. There is no way to define exactly the thickness of the atmosphere. The density of air diminishes with height, until at several hundred miles above the surface the atmosphere merges with the vacuum of space. Ten miles above the surface the density of the atmosphere has dropped to 1/10th of the surface value. Twenty miles up, the air is only 1/100th as dense as at the surface. Go up twenty miles and you have 98 percent of the atmosphere below you. On my drawing I have peeled up a piece of that twenty-mile-thick layer, like the skin of a potato, to show you how thin our umbrella of air really is.

Most meteorites begin burning up at heights of about 60 miles. Very few meteorites penetrate below 40 miles. But every now and then you will read in the papers about a fist-sized rock that came sailing through someone's roof. At even rarer intervals the Earth is struck by an asteroid that will knock a good-sized hole in the ground. These rare collisions of massive bodies with the Earth can have catastrophic consequences for life (see Chapter 15).

A more subtle and persistent rain from space, day and night, equator and poles, are the cosmic rays. Cosmic rays are charged subatomic particles traveling at nearly the speed of light. They approach the Earth from every direction, a kind of random high-energy atomic cloudburst. The sources of cosmic rays lie far beyond the Solar System. They include flares on distant stars, supernovas, explosions in the nucleus of our Galaxy, and violence in galaxies beyond the Milky Way.

Only the most energetic cosmic rays penetrate the atmosphere and reach the Earth's surface. Most cosmic rays detected at the surface are part of a shower of secondary particles created when a cosmic ray from space collides with an air molecule.

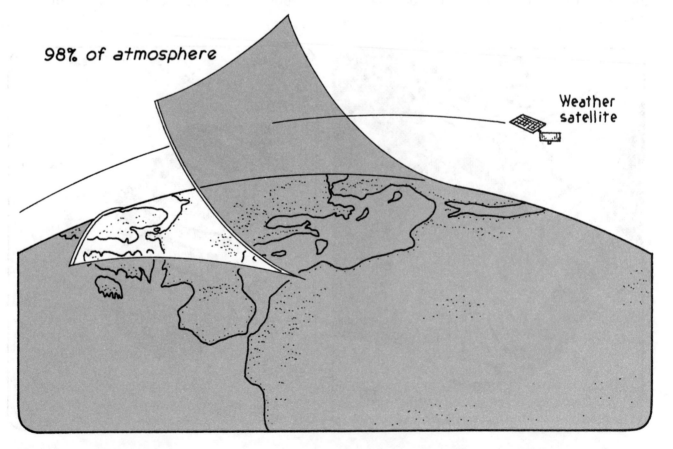

98% of atmosphere

Weather satellite

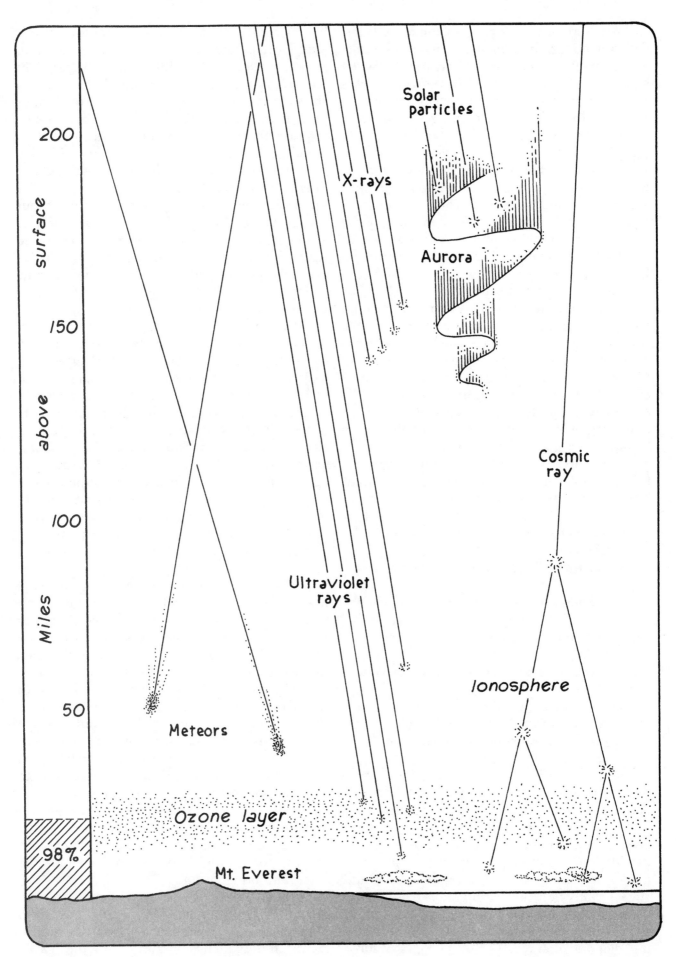

200

surface

150

above

100

Miles

50

98%

Solar
particles

X-rays

Aurora

Cosmic
ray

Ultraviolet
rays

Ionosphere

Meteors

Ozone layer

Mt. Everest

Since cosmic rays are subatomic particles, we have no sensation of injury if struck by one. The damage done, if any, is on the molecular scale. It is generally conceded that cosmic rays are an important source of gene mutations. Without the modest disrupting influence of cosmic rays it is likely that evolution would have proceeded at a more leisurely pace.

The Sun is the source of most of the radiations that fall upon the Earth. Without the Sun's radiation no life could exist. The Sun's infrared and visible light warm the Earth and supply the energy that sustains photosynthesis. Fortunately, the atmosphere is transparent to radiant heat and visible light.

But other solar radiations are less felicitous.

For one thing, there is a wind of charged subatomic particles that streams out from the Sun, particles like cosmic rays but generally less energetic. These solar particles are deflected and trapped by the Earth's magnetic field (see next chapter). Where they do enter the Earth's atmosphere, near the north and south magnetic poles, they collide with molecules of air to produce the beautiful auroral lights.

The Sun is a powerful source of x-rays. Artificially produced x-rays can serve an important function in medicine, but frequent exposure to x-rays is not recommended for living things. Fortunately, the Sun's x-radiation is absorbed by the rarefied air of the upper atmosphere.

Which brings us at last to ultraviolet light.

The Sun's ultraviolet radiation floods the planet, but a layer of ozone in the upper atmosphere prevents the deadly flux from reaching the surface. And a good thing, too. The sensitive nucleic acids in living cells, which preserve and transmit the genetic codes, are particularly susceptible to damage by ultraviolet radiation. Proteins are also at risk. Without protection against the Sun's ultraviolet radiation, life on Earth would be in constant jeopardy.

But by a curious paradox, if the atmosphere had been opaque to ultraviolet light from the very beginning, there might not have been a beginning for life on Earth. The flood of high-energy light on the early Earth hastened the production of organic compounds, the phosphates and amino acids and organic bases and sugars that were the building blocks of life and its first food supply. Ultraviolet light can tie chemical knots, but it can also untie them. Ultraviolet radiation was the tricky witch that lured life forward with sugary sweets and then struck with mortal effect.

Early fermenting life forms sheltered in dark places, deep water, or muddy bottoms, in clays or shoreline crevices, feeding on sweets produced by the Sun's ultraviolet light and hiding from the light itself. But the food supply was limited and there was no way for organisms to make their own food by photosynthesis without direct access to visible sunlight. This was the problem: how to soak up the good rays without being killed by the bad. Life devised a variety of strategems to solve the problem of coexistence with ultraviolet radiation, most of which depended on life under water. At moderate depths water screens ultraviolet while passing useful visible light.

Ultimately, life provided its own ultraviolet screen. Photosynthesis produced oxygen as a byproduct. Ultraviolet light can break an oxygen molecule apart into two atoms of oxygen, one of which can recombine with a two-atom molecule to form ozone. Ozone is unstable, but at heights of about 20 miles above the surface of the Earth a steady level of ozone is maintained by incoming ultraviolet radiation. Ozone absorbs ultraviolet light. When the oxygen content in the atmosphere approached present levels the ozone screen became effective. Living creatures were able to come out of the water or dark nooks and crannies and become full citizens of the land.

The first fossil evidence for a vigorous plant life on land dates from Devonian times, about 400 million years ago. I have shown *Archaeopteris*, a common fernlike fossil from Devonian rocks in New York State. The plant is also well known from the Old Red Sandstone of Ireland.

The ancestors of *Archaeopteris* had to solve many problems before pulling their feet out of the water: how to prevent drying out, how to hold themselves erect without the support of water, how to reproduce without relying on water to bring sperm and egg together, how to extract oxygen directly from the air.

The solutions to these problems were ingenious and various, and probably involved a long apprenticeship on shores and streams, in shallow ponds and lagoons, with one foot in the water and one foot out.

The evolution of successful land plants required the development of a vascular system to pump water and nutrients up from the soil to leaves and branches. Coatings on the leaves were needed that were transparent to oxygen and sunlight but kept precious water from escaping. And elaborate root systems were necessary for moisture, nourishment, and vertical stability.

It took millions of years for these innovations to be devised and tested. By Devonian times—400–350 million years ago—the solutions were well in hand and quite suddenly the world was green. Forests marched up from the shores. By a kind of cosmic magic, and maybe a few cosmic rays, seaweeds had transformed themselves into towering trees.

Not long after the plants successfully colonized the land, animals followed. The first land animals were amphibians, transient visitors to the dry shores that returned to the sea for reproduction. During Devonian times the backboned fishes ruled the sea, claiming the sceptre from the waning trilobites and sea scorpions. The Devonian period is sometimes called the Age of Fishes. The fishes were the ancestors of the amphibians.

In my drawing of a Devonian shorescape (next page) I have included two lobe-finned fishes clambering onto the shore. I fear that in my sketch they look rather like tourists on holi-day. Be assured that these creatures left the water only of necessity, flip-flopping across the muddy banks in search of food or a better water hole. The journey was almost certainly perilous. We can imagine these lobe-finned fishes gulping air and dragging themselves laboriously up from the water with fins better suited for swimming than walking. But the skills these early amphibians acquired over millions of years were not lost. Fins became feet, lungs evolved, skeletons became modified for life on land. By the end of the Devonian, amphibians were scampering among the trees and horsetails and ferns. Insects, mites, spiders, worms, and snails made the transition at about the same time. A taste for insects may have encouraged the fishes to seek a life on the sandy shore.

So the bleak moonscape of the Earth's dry lands was softened and the rocks came to life. Three hundred and fifty million years ago the Earth began to look familiar. A photograph from space would have shown the same blue-green planet that you and I have come to know and love.

I have on my desk a 16-inch Earth globe. If I dunked the globe in water and shook it dry, a

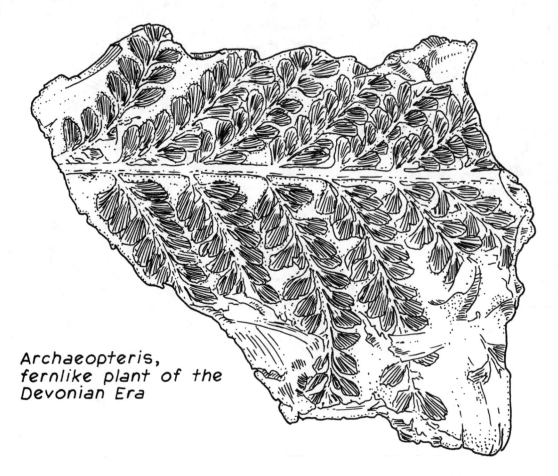

Archaeopteris, fernlike plant of the Devonian Era

Devonian Shorescape

thin film of moisture would cling to the surface. That film would be more than thick enough to represent the Earth's seas. Above the film of moisture imagine another film of air, no thicker than this letter *i*. In that double glaze of water and air—or rather, near the interface of that double glaze—life evolved and flourished.

The Earth spins with its thin liquid-gas envelope in black space, a tiny spherical stone hung to a star by a gravitational thread. The star is the source of the planet's life, but the star is also an abiding source of danger.

Only the atmosphere protects life on Earth from a rain of death from the sky. Only ozone protects life in the open from gene-breaking ultraviolet rays. There are various ways the careless use of human technology might endanger the precious ozone screen, and expose life on Earth to ultraviolet light. Caution is in order.

A Wind
from the Sun

The Earth's magnetic field deflects a wind of particles that flows from the Sun. On occasion the field collapses and the planet is left unprotected, with dire consequences for life.

On the evening of September 30, 1961, observers in the northeastern United States were treated to a spectacular display of the northern lights—the aurora borealis. The display was the first I witnessed (I grew up in the south) and remains the best show of the lights I have seen. It was a night that has lingered in my memory.

The display of lights against the dark night sky went on for hours, peaking, as I recall, about 10 P.M. I stood with friends on the roof of the physics building at the University of Notre Dame and gaped with wide-eyed wonder at the ongoing show. I had read, of course, about the lights, but I was unprepared for their dynamic beauty.

I have sketched an impression of the display from memory (do you recognize the Big Dipper?). The northwestern horizon was hung with shimmering green curtains of light. This celestial drapery wavered and folded upon itself as if moved by a gentle wind. Near the zenith a 4th of July fireworks show was under way. Streamers and starbursts of red and green light streaked the sky. Such vivid colors against the black night were totally unexpected and treated

by one and all in our group of elated watchers with gasps of disbelief.

The source of all these splendid pyrotechnics was the Sun. Two days earlier a massive flare had been recorded on the face of the Sun by workers at solar observatories. Within minutes a dark filament embedded in the face of the Sun had exploded into blazing sheets of flame, looping up into the Sun's atmosphere and tearing away into space. The flare blasted into space an enormous flux of charged particles—protons and electrons mostly—with velocities over a million miles per hour. Two days later a wave of those particles collided with the Earth. They were deflected by the Earth's magnetic field toward northern and southern latitudes. Smashing into atoms and molecules of the upper atmosphere, the concentrated stream of particles caused the air to glow. Typically, auroral lights are excited at an altitude of about a hundred miles.

Position of flare on face of sun

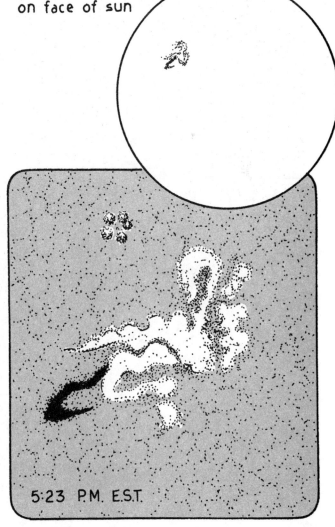

Solar Flare - September 28, 1961

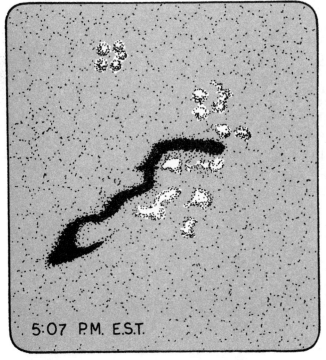

5:07 P.M. E.S.T.

5:23 P.M. E.S.T.

It was a particularly violent solar flare that excited the brilliant aurora of September 30, 1961. But even in less violent moments the Sun throws off a stream of subatomic particles, the so-called solar wind. Life on the surface of the Earth is protected against this potentially harmful breeze by the atmosphere and the planet's magnetic field.

The Earth's magnetic field has its origin in electrical currents that flow deep in the planet's liquid core. The field is similar to the field of an ordinary bar magnet, except that it is altered in shape by interaction with the solar wind. On the side toward the Sun the field is squeezed toward the Earth. On the side away from the Sun the field is stretched out like taffy.

Most of the wind of charged particles from the Sun is deflected by the Earth's magnetic field and flows harmlessly around the planet like water flows around a stone in a stream. Some of the particles penetrate the field and are trapped

in doughnut-shaped zones called the Van Allen radiation belts. Other of the more energetic particles penetrate the magnetic shield and stream in along lines of magnetic forces to collide with the atmosphere in northern and southern polar regions. Almost all of these successful intruders are stopped by the atmosphere, and in stopping ignite the auroral lights. Only the most energetic particles reach the surface of the Earth, and most of these come to ground in polar regions where life is sparse.

Very intense storms on the Sun throw superfast particles into space that are able to penetrate the Earth's magnetic field at low latitudes. It is these superfast particles that produce occasional light shows for observers in temperate zones. The lights are beautiful. They are also a healthy sign that the atmosphere is stopping those bits of the solar wind that are not magnetically deflected.

Several decades ago earth scientists became

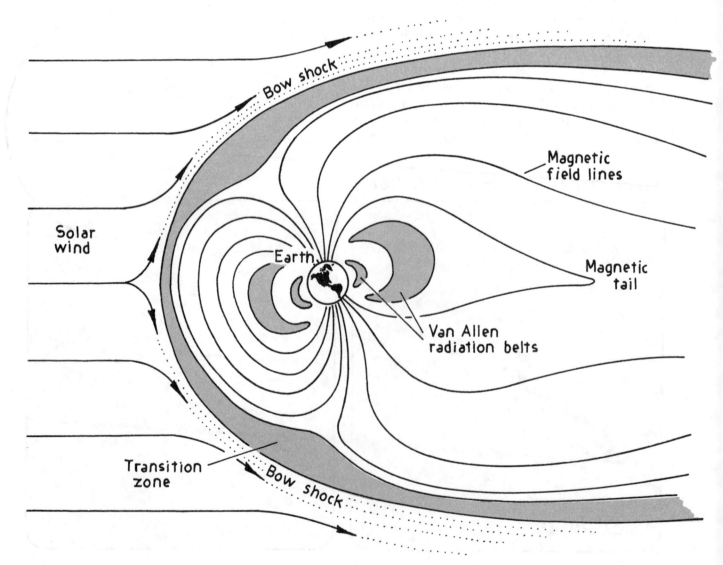

aware of an extraordinary and unexpected feature of the Earth's magnetic field: at unpredictable intervals, typically half a million years, the field reverses. North magnetic pole becomes south pole and vice versa. During a reversal, the needle of a compass would swing around and point the other way!

Geophysicists do not yet understand what causes these bizarre flip-flops of the Earth's field. They can, however, reconstruct past reversals by using fossil magnetism in the rocks. I have illustrated below a detailed schedule of field reversals over the past 4 million years, and in a less exact way the general behavior of the field over the past half-billion years. You will notice that the field configuration labeled "normal" is really no more common than the intervals labeled "reversed." "Normal" means no more than the polarity we are used to today.

During the past tens of millions of years the field has been flip-flopping with reliable fre-

quency. As the bottom graph shows, there have been long eras of Earth history when the field has remained rather steadily in one polarity or the other. Since no one understands with certainty why the field flips in the first place, the intervals of stability are equally mysterious.

By examining the buried fossil magnetism in sea-floor sediments, earth scientists have reconstructed the detailed behavior of the field during a flip-flop. The turnabouts are not instantaneous. The field slowly collapses from one configuration and builds up again with opposite polarity. A reversal may take ten or twenty thousand years. For a period of several thousand years the planet is effectively without a magnetic field!

Upon learning of this latter fact, scientists began speculating about the effect of reversals on living things. If there are long intervals when the Earth is magnetically naked before the solar wind and cosmic rays from space, might not

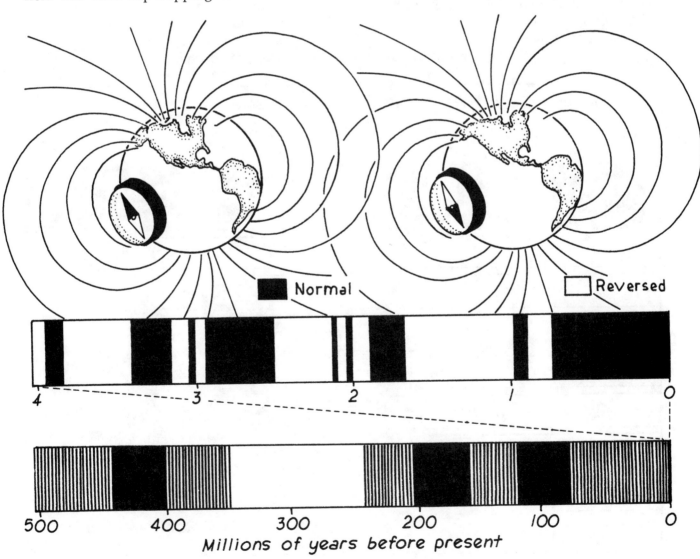

Normal   Reversed

4   3   2   1   0

500   400   300   200   100   0

Millions of years before present

plants and animals feel the brunt of the celestial bombardment?

During the late 1960s and early 1970s several studies revealed tantalizing evidence that polarity reversals might indeed effect the course of evolution. I have illustrated the results of just one study, carried out by J. D. Hays of the Lamont-Doherty Geological Observatory. Hays studied radiolaria in 28 drill samples of sea-floor sediments (called cores) brought up from the floor of the Pacific. Radiolaria are one-celled marine organisms that secrete delicate and beautifully filigreed skeletons of silica. Hays found eight species of radiolaria that became extinct within the past 2.5 million years. Of the eight, six species disappeared from the sediments at the time of polarity reversals. Other studies of microorganisms in sea-floor sediments have confirmed and extended Hays' work.

At least five mechanisms have been proposed to explain how the collapse of the Earth's magnetic umbrella could lead to disruptions in the evolutionary record.

The most obvious possibility is a direct bombardment of exposed organisms by the solar wind and cosmic rays. The bombardment might produce genetic damage that leads to the extinction of a species. This turns out to be unlikely. Even in the absence of the magnetic field, the Earth's atmosphere should provide a reasonably effective shield against incoming particles.

A second proposal links polarity reversals with geological activity in the Earth's upper mantle. According to this idea, whatever forces trigger periods of intense earthquake and volcanic activity might also be the trigger for a polarity change. Dust thrown into the atmosphere by volcanic eruptions leads to climatic change, which in turn causes evolutionary chaos. In this scenario, the reversals are not the cause of extinctions, but a coincident effect. Evidence has been offered to show that some past episodes of intense volcanic activity were in fact coincident with reversals.

There are ways in which a polarity flip-flop might directly cause a change in climate. During the time when the Earth is without its magnetic sheath, incoming solar particles and cosmic rays increase ionization (atoms stripped of electrons) in the upper atmosphere. Ionization seems to

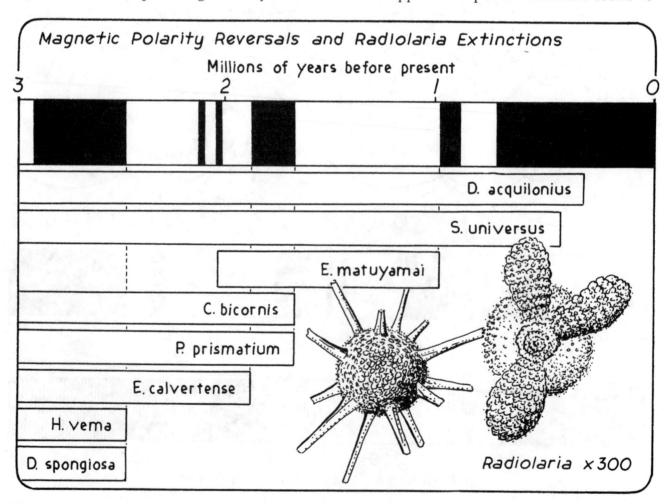

Magnetic Polarity Reversals and Radiolaria Extinctions

Millions of years before present

3   2   1   0

D. acquilonius

S. universus

E. matuyamai

C. bicornis

P. prismatium

E. calvertense

H. vema

D. spongiosa

Radiolaria x300

enhance high-altitude cirrus cloud cover. Cloud cover modifies climate. According to this third theory, changes in the upper atmosphere during a polarity reversal modify the weather sufficiently to cause the extinction of certain life forms.

A fourth theory relates the collapse of the Earth's magnetic shield to a depletion of the ozone layer. An increase in the level of ionization in the stratosphere encourages a reaction of nitrogen and oxygen which leads to the formation of large quantities of nitric oxide. Nitric oxide in turn speeds the destruction of ozone by reaction with the ozone molecule. Diminishing ozone exposes the surface of the Earth to potentially dangerous ultraviolet radiation from the Sun. The effect would be especially great during periods of intense solar activity.

Those species of microorganisms which live near the surface of the sea and which evolved over a long period of magnetic stability might be susceptible to the harsher ultraviolet environment that occurs during a reversal. This would be especially true for creatures which already occupy precarious ecological niches.

A last explanation for extinctions during magnetic reversals takes no account of the solar wind or cosmic rays. The source of the effect is closer to home, within the bodies of living things.

In recent years substances responsive to magnetic fields have been found in creatures as diverse as bacteria, molluscs, mud snails, honeybees, butterflies, pigeons, and dolphins. These internal "compasses" seem to play a role in navigation. A flip of the field could cause dangerous disorientations. Bacteria, for example, might find themselves swimming toward the surface of a pond rather than toward the muddy bottom.

Life has evolved in a magnetic field and may be more sensitive to changes in the field than we have previously suspected. If this is so, some forms of life may also be put at risk by changes in the Earth's magnetic field. It is a subject that requires further study.

Those creatures that evolved during long periods of magnetic stability may be particularly at risk when the field begins flipping. There are hints in the fossil record that some past episodes of faunal extinction may be related to these crucial flip-flop moments in the Earth's magnetic history.

One thing is clear; when the Earth is without its magnetic shield, life is in jeopardy. The solar wind falls full-face upon the planet with potentially dire consequences.

The Earth's magnetic field may also provide some protection against another potential source of high energy particles—the violent deaths of nearby stars. More on this in the next chapter.

Upon the Deaths of Stars

A nearby supernova would flood the Earth with high-energy radiation, modifying the atmosphere and the climate. Such events may have instigated crises in the history of life.

Late one Friday night in August 1975, I received a telephone call from a friend. He had just heard a radio report of a new star in the constellation Cygnus the Swan. I rushed outside to the scene at left. It took but an instant to recognize the intruder.

This part of the sky is one of the most brilliant for northern observers. Here can be found the fullest stream of the Milky Way, split from top to bottom by the Great Rift. And here are the three first-magnitude stars—Vega, Deneb, and Altair—that make up the famous "Summer Triangle."

Deneb is the Arabic word for "tail." The star represents the fanned tail of a long-necked swan that wings its way south along the stream of the Milky Way. On that particular August evening it looked as if a tail feather had been plucked from the swan as it flew by. Not far from Deneb, where no star had been before, was a new star almost as bright as Deneb.

The new star was short-lived. By the following evening it had dimmed noticeably. Within a week the star was visible only to observers with telescopes or binoculars.

Nova means "new," but a nova is not really a new star. What we see is the sudden brightening of an old star that was too far away to be ordinarily visible. The star that brightened in August of 1975 was thousands of light-years away, all the way across one of the spiral arms of the Milky Way Galaxy, beyond the reach of the largest telescopes. The flaring of a nova occurs at the end of a star's life when the energy balance within the star becomes unsteady and the star blows off its outer layers.

There is a beautiful example of an ex-nova in the same part of the sky as the nova of 1975. In the constellation Lyra, near gorgeous Vega, an observer with a telescope can find the Ring Nebula. The amateur telescope shows a wispy wreath of smoke. Observatory photographs reveal the tiny central star—or skeleton of a star—and the dazzling spectral colors of the expanding shell of gas.

The cause of novas such as Nova Cygni 1975 are violent convulsions that occur late in the lives of sunlike stars. Stars more massive than

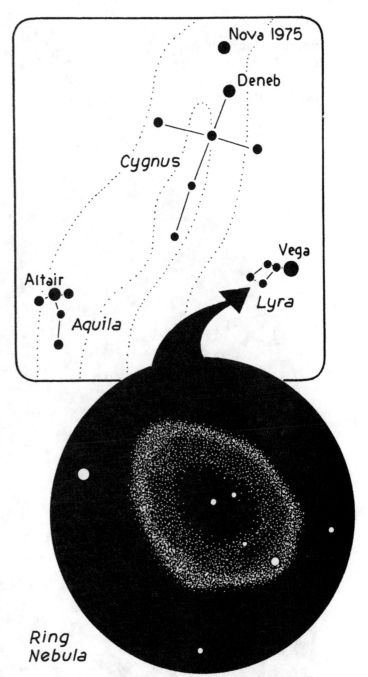

Ring Nebula

the Sun, the giants of the Galaxy, depart this world with even greater violence. The superstars blow themselves to bits in cataclysms called supernovas. Supernovas pour into space enormous shock waves of matter and energy. Woe to the unlucky planet that finds itself in the way.

There are hundreds of billions of stars in the Milky Way Galaxy. There have been four supernovas on our side of the Galaxy during the past 1000 years, including the Crab supernova of A.D. 1054. None of the four stars that self-destructed was close enough to pose a threat for life on Earth.

What are the chances that a relatively nearby star will go supernova, a star close enough to threaten life in its passing?

On the map at right I have shown all stars brighter than the Sun that lie within a 50 light-year radius of the Sun. This is a flat map, drawn in the plane of the Milky Way; the stars are actually distributed in three dimensions. This 50 light-year neighborhood is only a tiny corner of

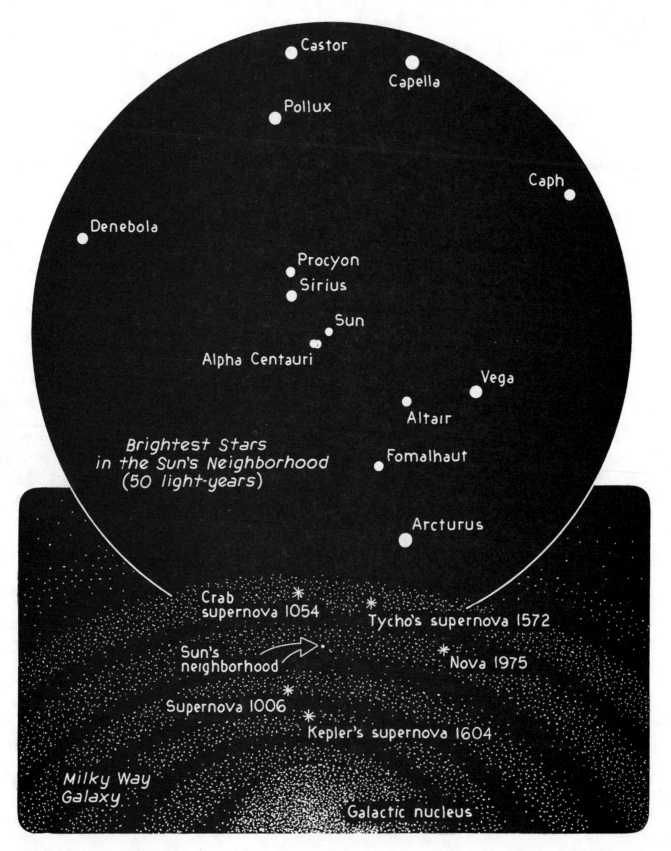

Castor

Capella

Pollux

Caph

Denebola

Procyon
Sirius

Sun

Alpha Centauri

Vega

Altair

Fomalhaut

Brightest Stars
in the Sun's Neighborhood
(50 light-years)

Arcturus

Crab
supernova 1054

Tycho's supernova 1572

Sun's
neighborhood

Nova 1975

Supernova 1006

Kepler's supernova 1604

Milky Way
Galaxy

Galactic nucleus

the Galaxy, no bigger than the dot on my drawing of our side of the Milky Way Galaxy.

Within our neighborhood, the Sun is one of the larger stars. Only thirteen stars in the neighborhood are more luminous than the Sun. A complete census of the neighborhood would include a thousand additional stars, many so small and dim that they have not yet been observed. These tiny red and orange dwarf stars are not candidates for supernovas. Their deaths will be orderly and gentle—if and when they die. Their lifetimes are likely to be very long. It is the stars larger than the Sun that are potential threats.

The more massive a star, the shorter its lifetime. The Sun should have a lifetime of about 10 billion years. A star like Vega will live for only several hundred million years, a small fraction of the time that life has existed on Earth. Capella, Pollux, and Altair are approaching the end of their lives and have swollen to become orange giants. But none of these stars are the super massive giants that might pop off with supernova violence. Deneb in the swan's tail is a good candidate for a supernova, but Deneb is a safe thousand light-years distant, far beyond the Sun's neighborhood.

Stars have independent motions in space, and over millions of years some of our neighboring stars will move away to be replaced by new neighbors. Meanwhile, the Sun travels through the Galaxy. All of this moving around complicates any attempt to estimate the chances of a nearby cataclysm. A reasonable estimate is that a supernova will occur within our 50 light-year neighborhood once every few hundred million years.

Even if this guess is off by a factor of ten, it is still worth considering the effects of a nearby supernova for life on Earth. If a supernova occurred at the distance of Vega, the Solar System would lie within the shell of exploded debris for hundreds of years. The shells of supernovas carry trapped cosmic rays. A nearby supernova explosion might subject the Earth to a cosmic ray bombardment hundreds of times greater than normal for long periods of time.

In addition, the explosion of a nearby star would sweep the Earth with a shorter burst (days or weeks in duration) of gamma rays, x-rays, and a straggling wave of high energy cosmic rays moving at nearly the speed of light.

Many researchers have attempted to access the biological effects of this flux of radiation—with widely diverse results. No one predicts that a supernova would take life down for the count, but almost everyone who has studied the problem agrees that the blast of cosmic fire resulting from a nearby supernova would be a staggering blow.

The effect of the blast could take several forms. The first would be an increase of the radiation that penetrates to the surface of the Earth, with direct molecular consequences for life. Plants appear to be hardier than animals in the face of radiation. Denizens of deep waters would be protected, but they might be susceptible to disruptions of food chains that rely on surface organisms. The cells of living creatures have the ability to repair some molecular damage. Whether the repair service would be swamped with calls following a nearby supernova remains to be seen.

A more persistent threat to life might be depletion of the ozone layer, by the same mechanism discussed in the previous chapter. A supernova within 50 light-years of the Earth would cause atmospheric ionization hundreds of times greater than normal. The resulting ozone depletion could be 50 percent or more for periods of years or centuries. Unscreened solar ultraviolet radiation would have its way untying the delicate knots that hold together the threads of life. Nocturnal animals would be safer than those that go abroad by day.

Increased levels of ionization would certainly modify climate. Most researchers calculate a cooling trend, perhaps great enough to induce glaciation. A few have been bold enough to suggest that supernovas were the initiators of the great ice ages of the past.

If one of the stars in our neighborhood went supernova, for a time it would shine as brightly as the full moon. You could read by its light. Of course, at a distance of 50 light-years we would not see or feel the effects of the star's self-destruction until 50 years after the actual detonation. Just as the brilliant new object appeared in our sky we would be hit by a first wave of destruction, gamma rays, x-rays, and high energy cosmic ray particles. Several thousand years later the shell of exploded material would arrive, still wreathed in deadly radiation. The death of the star would deliver a one-two punch which life would have to absorb.

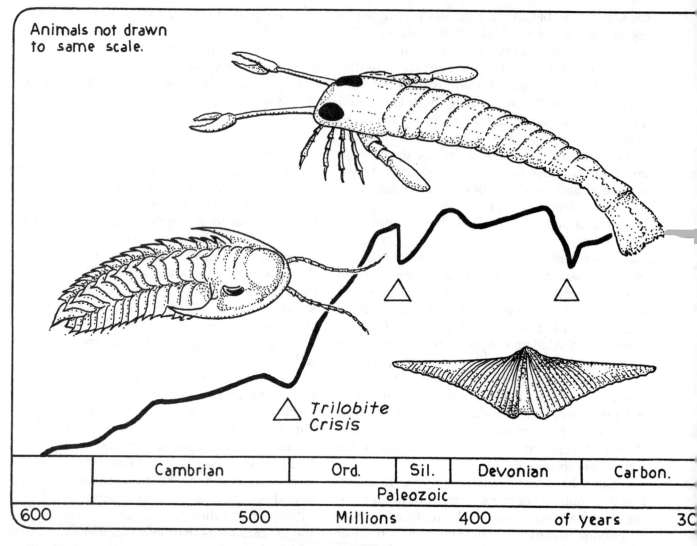

Animals not drawn to same scale.

△ Trilobite Crisis

	Cambrian	Ord.	Sil.	Devonian	Carbon.
			Paleozoic		

600   500   Millions   400   of years   3C

Is there any evidence that life has stumbled beneath such blows in the geological past? There is certainly evidence of sharp crises in the history of life, although none that ties a crisis directly to a supernova.

I have illustrated above the results of one recent survey of the fossil record. The line of the graph represents the standing number of families of marine animals—vertebrates, invertebrates, and protozoans—over geologic time as recorded in the fossil record. The study encompassed 3300 families, of which 2400 are now extinct. The general upward slope of the graph is a net gain of diversity over the background of extinctions that goes on all the time.

What is most interesting about the graph is the handful of episodes in the history of life when the rate of extinctions suddenly soars and the number of marine families takes a sharp dip. The most spectacular of these is the so-called "Time of the Great Dying" which occurred 230 million years ago. Nearly half of all animal families disappeared from the Earth. Seventy-five percent of the amphibians and eighty percent of the reptiles were wiped out. Worst hit were creatures that lived in the sea, especially the invertebrates. Corals, crinoids, blastoids, ammonoids, brachiopods, bryozoans, molluscs, formaminifera, and fishes were devastated. The trilobites, which had suffered several earlier disruptions, were finally pushed into oblivion.

So abrupt and dramatic were these episodes of extinction that the discontinuities in the fossil record have long been used by geologists to define periods and eras of geological time. The classical breaks between the Paleozoic ("old life") Era, the Mesozoic ("middle life") Era, and the Cenozoic ("recent life") Era were defined by the two greatest crises in the history of life.

The cause or causes of these breaks in the record of life are unknown. As we shall see in the next chapter, there has recently been intense speculation about the cause of the most recent

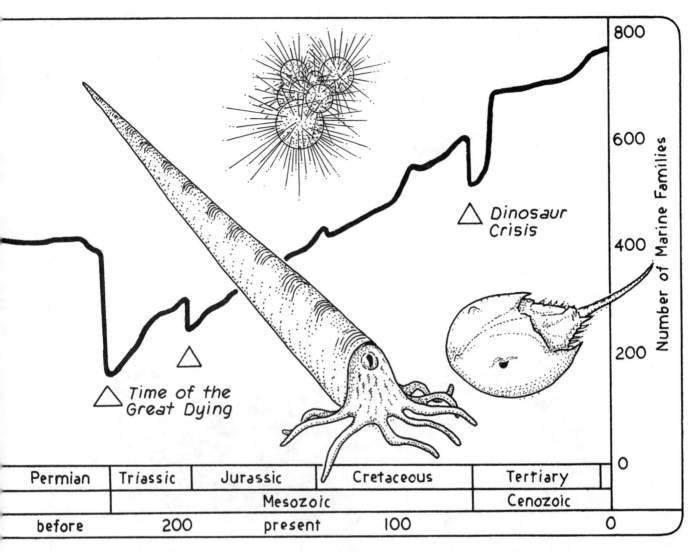

Permian	Triassic	Jurassic	Cretaceous	Tertiary
		Mesozoic		Cenozoic
before	200	present	100	0

Number of Marine Families

Time of the Great Dying

Dinosaur Crisis

mass extinction, the one that terminated the reign of the dinosaurs. The causes of the earlier episodes, including the "Time of the Great Dying," are up for grabs.

Supernova events must certainly be considered as possible causes for mass extinctions. But they are not the only candidates for that dishonor. Magnetic polarity reversals, asteroid bombardment, volcanic activity, changing sea levels, and climatic change have all been cast in the role of assassin.

Perhaps no single cause was responsible for the episodes of extinction. At times of severe geological stress, life may be more vulnerable to punctuational events like supernovas or magnetic polarity reversals. The explosion of a nearby star or the flip of the Earth's magnetic field may simply be the straw that breaks the camel's—or dinosaur's, or trilobite's—back.

But never underestimate the resiliency of life. The definition of life *is* survival. The death of one individual or species is compensated by the genesis of another. And so the curve of diversity climbs in spite of temporary setbacks. Nor is life a passive plaything of the environment. Pressed by unhappy circumstance, life has shown itself capable of changing the environment to its own liking.

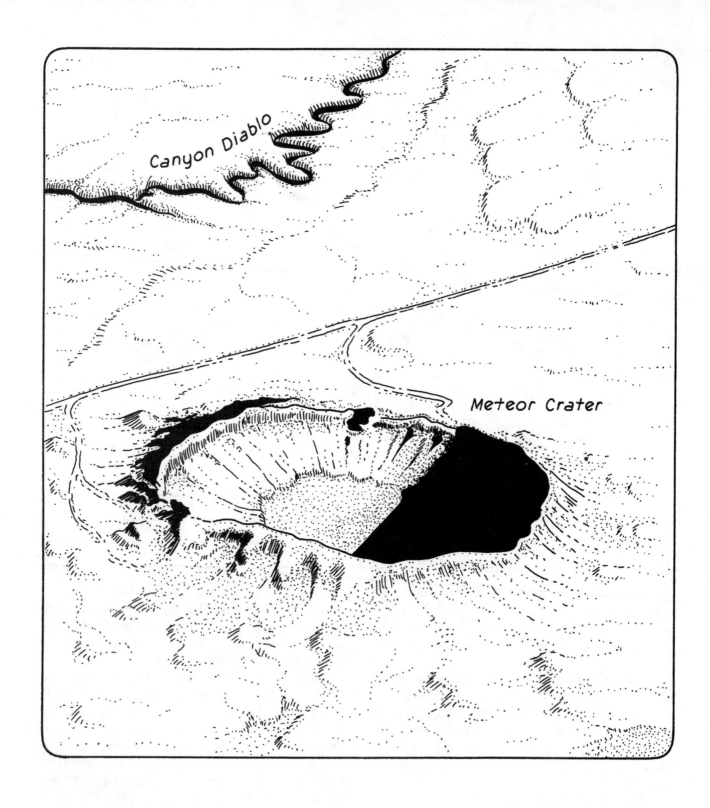

Canyon Diablo

Meteor Crater

Asteroid and Dinosaur

What killed the dinosaurs? Evidence has accumulated that the culprit was a large Earth-impacting asteroid, thereby sparking a lively controversy.

The meteor was traveling through space at tens of thousands of miles per hour when it collided with the Earth. It was as big as a house and weighed a hundred thousand tons. It blasted a hole in the ground nearly a mile in diameter.

Today, the crater in the desert near Flagstaff, Arizona is a popular tourist attraction. It is the newest impact scar on the face of the Earth, and a vivid reminder of the danger that can come from the sky.

The Arizona meteor crashed to Earth 25,000 years ago, at about the time the first humans entered North America from Asia. If there were witnesses to the event, they saw a demonstration of celestial violence that has not been equalled since.

If a similar object slammed into Arizona today, the human toll might be staggering. If the object fell into an ocean it would raise a tidal wave that would devastate coastal cities. It is a sobering thought to realize that such an event is not only possible, but—given enough time—certain.

In recent years geologists have made a close census of ancient impact craters on the Earth, Moon, and other planets. Astronomers have studied those asteroid-sized bodies that swing through the Solar System on Earth-intersecting trajectories. Together they conclude that several objects capable of excavating a crater at least 6 miles wide will crash into the Earth every million years. The ecological consequences of such an impact, on land or sea, would be enormous. Could asteroid impacts have punctuated the evolution of life on Earth?

Some scientists say yes.

In particular, an Earth-colliding asteroid has been championed as the cause of the wave of plant and animal extinctions that swept the Earth 63 million years ago. The extinctions are marked by a sharp discontinuity of life forms preserved as fossils in the stratified rocks. The discontinuity defines the boundary between the Cretaceous and Tertiary periods of geologic time, and between the Age of Reptiles and the Age of Mammals. The most famous victims of the Cretaceous-Tertiary calamity were the dinosaurs.

One place where the boundary between the two geologic eras is exposed for inspection is in the hills near Gubbio, Italy. A thin layer of clay separates two formations of marine limestone. The limestone below the clay contains fossil marine organisms typical of Cretaceous times. There are no fossils in the clay. Tertiary fossils characterize the limestone above the clay.

In 1979 a group of scientists at the Lawrence Berkeley Laboratory in California announced the discovery of an abnormally large concentration of the rare element iridium in the clay layer at Gubbio. The iridium level was 30 times greater than in the rocks above or below the clay.

Iridium is rare in rocks of the Earth's crust, but it is found with much richer abundances in meteorites. This led the Berkeley group to suggest that the clay layer at Gubbio was deposited on the sea floor following the collision of a massive asteroid with the Earth. The clay presumably was the fallout of a mixture of meteoric material and pulverized Earth rock blasted into the atmosphere by the impact.

In the months following the announcement by the Berkeley group, other investigators found exceptional iridium concentrations in marine sediments at sites as widespread as New Zealand, Denmark, and Texas. In each case the anomaly was associated with the Cretaceous-Tertiary boundary. The evidence mounted for a worldwide dusty fallout of exotic composition.

One of the most impressive iridium enhancements was reported in late 1981 for a site in the Raton basin of northeastern New Mexico. The sediments in the basin consist of layers of mudstone, sandstone, and coal that were deposited tens of millions of years ago in a freshwater swamp. The thousandfold iridium anomaly was associated with a thin layer of coal at the Cretaceous-Tertiary boundary. It coincided with an abrupt drop in the ratio of pollen grains to fern spores, and the extinction of several kinds of pollen. There is a simplified presentation of these discoveries at the bottom of this page (see Note).

Very little clay or other finely divided matter occurs in the coal at the New Mexico iridium anomaly, in contradiction to what might be expected if the Earth was blanketed by fallout. Nevertheless, the New Mexico study, and others like it, seemed to offer compelling support for the idea that the dinosaurs and companions were dealt a sharp, deadly blow by a cosmic intruder.

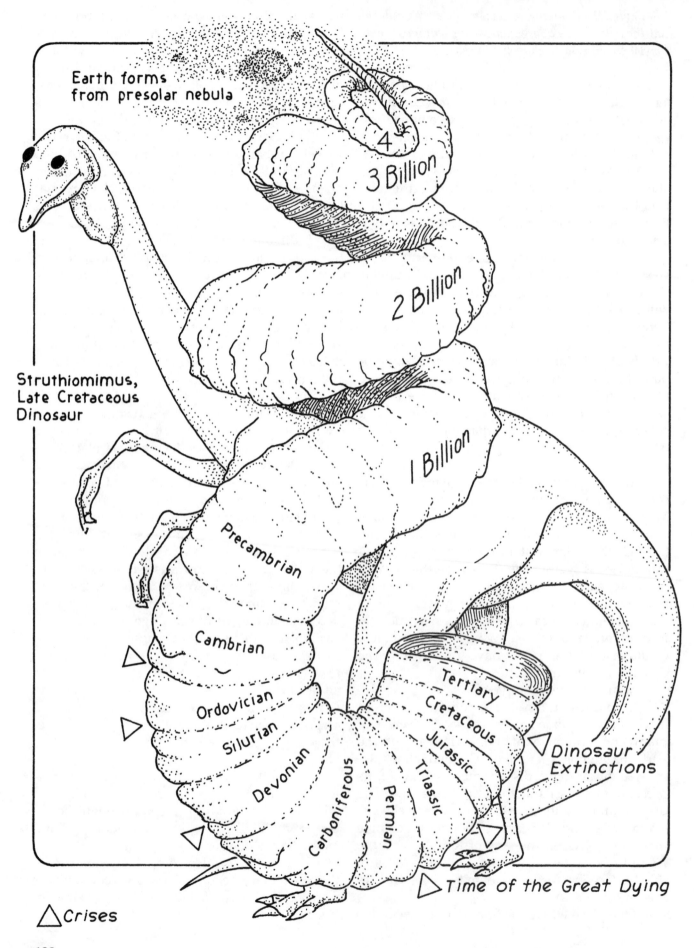

Earth forms
from presolar nebula

4

3 Billion

2 Billion

1 Billion

Struthiomimus,
Late Cretaceous
Dinosaur

Precambrian

Cambrian

Ordovician

Silurian

Devonian

Carboniferous

Permian

Triassic

Jurassic

Cretaceous

Tertiary

Dinosaur
Extinctions

Time of the Great Dying

△ Crises

108

Supporters of the asteroid hypothesis estimate that the impacting object had a diameter of 3 to 10 miles and released the energy equivalent of a hundred million hydrogen bombs. The consequences of such an event are difficult to assess, but a likely scenario follows.

The space invader passed through Earth's atmosphere and ocean like a bullet through tissue paper.

The impact excavated a crater the size of Rhode Island. If the impact was on land the shock wave alone spread terrible destruction. If the impact was in the ocean it raised a tidal wave three miles high. The wave washed over entire continental margins as if they were strips of sandy beach.

Perhaps as much as half of the energy carried by the asteroid was transferred to the atmosphere, giving rise to a short and potentially lethal heating pulse. Land animals were particularly vulnerable to this sudden rise of temperature. It appears that most land animals weighing

more than 50 pounds became extinct at the time in question.

At impact, a huge mass of material, rich in extraterrestial elements, was lofted into the atmosphere. Winds carried the debris worldwide, wrapping the Earth in a grey dusty shroud. For a year or more the Earth cooled, the oceans by a few degrees, the land areas by more. Some investigators have calculated a drop in land temperatures by as much as 20 degrees Fahrenheit.

For several months light levels at the surface of the Earth were below the limit of dark-adapted human vision. The entire planet was plunged into inky darkness. Animals, particularly large ones, had difficulty finding food.

For an even longer time photosynthesis was disrupted. It is thought that the temporary cessation of photosynthesis had particularly dangerous consequences in the oceans. The destruction of planktonic organisms in surface waters could have led to a collapse of food chains. The fossil record suggests that half of the

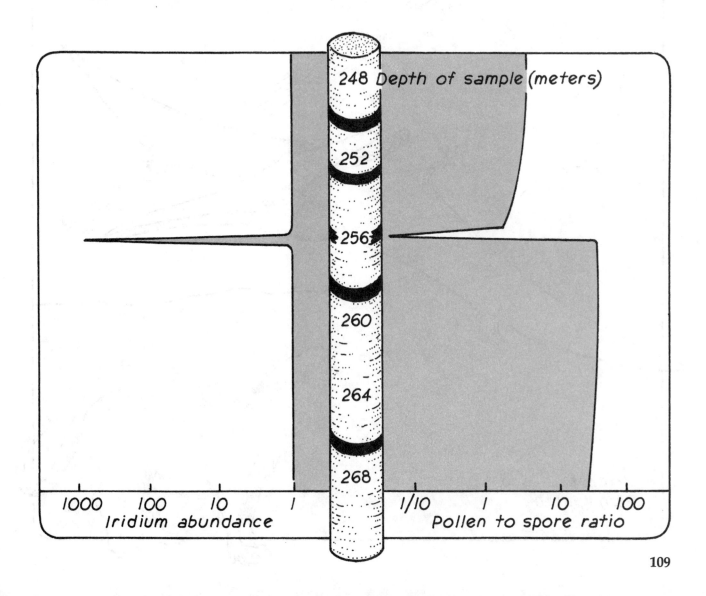

248 Depth of sample (meters)

252

256

260

264

268

1000   100   10   1
Iridium abundance

1/10   1   10   100
Pollen to spore ratio

floating marine organisms became extinct at about this time.

Of all the victims of the extinctions at or near the Cretaceous-Tertiary boundary, the most conspicuous were the dinosaurs. Dinosaurs had ruled the Earth for a hundred million years. They included the largest animals ever to walk the globe. Just prior to the time of the extinctions, familiar families of dinosaurs included the heavily armored plant-eating *Triceratops*, and the ferocious meat-eater *Tyrannosaurus*, the king of the race. As representative of the ill-fated reptiles I have illustrated a pterosaur, member of a family of flying reptiles that achieved wing spans of 50 feet! No larger animals ever took to wing.

The pterosaurs and their giant cousins fell before the asteroid catastrophe and yielded the Earth to more resilient creatures. The air was inherited by the birds, feathered fliers that somehow survived the cold dark aftermath of the asteroid collision. The land became the domain of the mammals, small agile creatures that had carved out a lifestyle scampering between the legs of the thunder-lizards. The burrowing nocturnal mammals fared the crisis particularly well. I have represented the mammals with condylarth, an early hooved ancestor of present grazing animals. As the reptiles slipped toward oblivion, the mammals raced toward dominance.

A few reptiles survived the late Cretaceous catastrophe. The turtles may have passed the crisis by staying under water. The ancestor of the crocodiles may have saved its species by its habit of burying its eggs in mud.

The asteroid version of the dinosaur extinc-

Pterosaur and
Condylarth

tions is guaranteed to appeal to the sort of mind that bends toward catastrophic explanations of the past. But it is not a "worlds-in-collision" fantasy. The story finds ample support in the record of the rocks. Between 1980 and 1983 the hypothesis gained increasing scientific acceptance. A NOVA program on the subject gave the theory a wide public following.

But not all scientists are willing to accept a cosmic intruder as the agent of extinction. It is not altogether clear, for example, that the "Cretaceous-Tertiary" iridium anomalies were in fact exactly coincident in geologic time. Many paleontologists argue that the fossil record supports a gradual and staggered wave of extinctions, rather than a single sudden catastrophe. Some maintain that the dinosaurs had disappeared from the fossil record well before the iridium layer that marks the supposed asteroid impact. And no one has pointed out a suitable candidate for the crater that the asteroid must certainly have blasted into the face of the Earth, although this need not be a damaging objection if the asteroid fell onto the floor of the sea.

A careful mineralogical study of the Cretaceous-Tertiary clay layers was potentially damaging to the asteroid hypothesis. If the clays were part of a worldwide fallout from an asteroid collision they should be mineralogically homogeneous from site to site. Second, the clays should include a terrestial component that is different from locally derived clays. Third, the clay should contain exotic minerals not normally found in marine sedimentary rocks.

Several investigators maintain that the clays differ considerably from place to place. They have tried to show that the clays are not widely different from local sediments above and below the boundary layer, and that there is no evidence of the kinds of exotic constituents that might be expected from an extraterrestrial source. If this is so, then the clays are more consistent with local volcanic sources than fallout from a cosmic intruder.

The Late Cretaceous era was a time of intense volcanic activity all around the Pacific. There were major regressions of the seas from the land. It was a time of changing climate. As we shall see in the next chapter, the crust of the Earth is a dynamic and often violent engine that can generate enough dangers for life without requiring the intervention of intruders from space.

Few scientific theories of recent times have excited more lively debate than the asteroid theory for the dinosaur extinctions. The controversy has been a healthy lesson in how science works. For a few years a growing tide of enthusiasm carried the asteroid theory forward. But the bandwagon was held in check by an ongoing critical reappraisal of the evidence and a careful testing of counter-hypotheses. The outcome of the story remains to be heard.

At the time of the Cretaceous-Tertiary extinctions some of the smaller carnivorous dinosaurs had achieved a brain-to-body weight equal to the early mammals. Some scientists have suggested that if the dinosaurs had not become extinct, these resourceful, intelligent reptiles would have retained their dominance on the planet. It is interesting to wonder if a big-brained scaly reptile might be writing this book today if an object hurtling through space had not collided with the Earth.

Continents
Adrift

The crust of the Earth is dynamic. Every few hundred million years the continents are rearranged and the ocean floors are made anew. Life must adapt its drama to the shifting stage.

Every rock, every pebble, every grain of sand has a story to tell of the Earth's past.

Here is an outcrop of bedrock at the base of a hill near my home in eastern Massachusetts. The surface of the outcrop is as smooth as glass and is marked with fine parallel scratches that run in a north-south direction. The polish and the striations are the work of the glacier that lay upon New England 10 thousand years ago. Here, on this outcrop, is compelling evidence for the Ice Age.

But it is not the Ice Age that is our business now, nor the scratches, nor the polish. It is the rock itself that attracts our attention. It is a fine-grained green volcanic rock (volcanoes in New England?). It is slashed across by wedge-shaped intrusions of coarse-grained pink granite. The green volcanic rock and the pink granite are igneous rocks, formed by the cooling and crystallization of molten minerals.

It is clear from the look of things that the granite wedged its way into preexisting volcanic rock. The volcanic rock must therefore be the older formation. Our geology books tell us that granite forms miles underground from the recrystallization of molten rock. But volcanic rock forms from lava on or near the surface of the Earth. We are forced to conclude that the volcanic rock made a trip down into the Earth and back again, picking up the intrusion of granite along the way.

Let's be more specific. First, in a period of volcanic activity, lava poured out onto the surface and solidified to form the fine-grained green rock. Then, somehow, the volcanic formation was carried or buried miles underground. In an episode of deep local heating, daggers of molten granite were squeezed into the green rock. The granite cooled. Then both formations were lifted toward the surface and the overlying layers of rock were eroded to reveal at last this particular piece of the Earth's crust. Hundreds of millions of years were surely required for the working out of all this geological violence.

My New England outcrop hints at a story of the wholesale restructuring of continents across eons of time, of continents flooded with the liquid marrow of the Earth, and of continents crumpled into high mountains that time levels. The theory of plate tectonics provides a satisfying mechanism for this global remaking of continents. The theory was created during the 1960s and is today almost universally accepted by geologists. Plate tectonics (known popularly and less accurately as "continental drift") gives geologists a unified understanding of most of the major features of the Earth's crust.

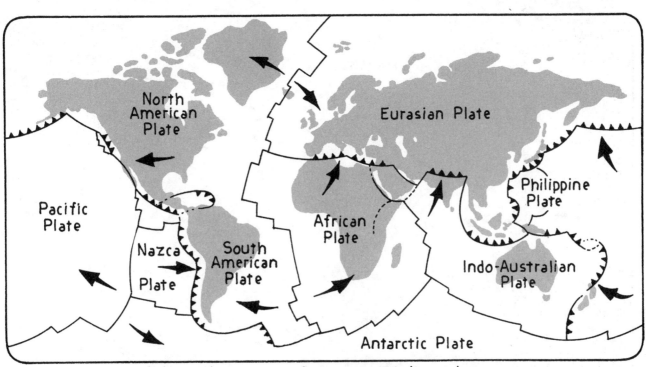

Divergent boundary   Convergent boundary

According to the new geology, the rigid crust of the Earth is as thin, relatively speaking, as the shell of an egg. It rests on the hot, nearly molten rock of the upper mantle. The "eggshell" is called the lithosphere, from the Greek for "stone." The hot plastic upper mantle is called the asthenosphere, from the Greek for "weak." Thermal and gravitational forces in the asthenosphere stress the rigid lithosphere and break it into sheets called plates. The crust of the Earth is like a cracked eggshell.

The plates move! The pieces of cracked eggshell slide about! That is the astonishing discovery at the heart of the new geology. I have illustrated the way the plates move with a side view of the Nazca Plate, part of the floor of the Pacific Ocean.

Along a system of mid-ocean ridges the plates move apart. As they separate, hot slushy rock oozes up from below to fill the gap. The lava freezes to form new sea-floor crust, which is again riffed and pulled apart. Again lava rises from below. In this way new eggshell for the Earth is continuously created along the ocean ridges.

But new crust cannot be created in one place without being consumed somewhere else. Along other plate boundaries the plates push together, and one of the colliding plates is forced down into the mantle. As the plate bends to take its dive, it creates an ocean trench. Almost all of the earthquakes and volcanoes on the Earth occur at plate boundaries. The boundaries are "where the action is."

The continents are made of lighter stuff than the floors of the oceans. They ride like passengers on the moving crustal plates. When a continent is carried to a convergent boundary, it resists being pulled into the mantle for the same reason a block of wood resists forcible submersion. Instead, the continent crumples and mountains are born. The great mountain ranges of the world were squeezed up by collisions of moving plates.

The push and shove of crustal plates has been going on for at least half a billion years, and

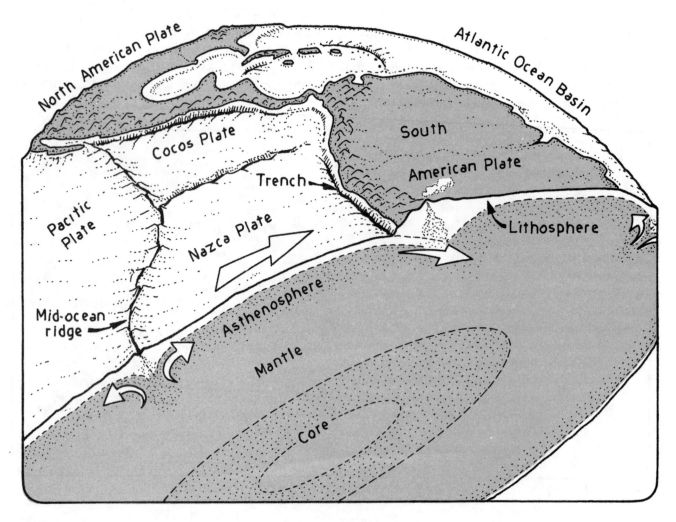

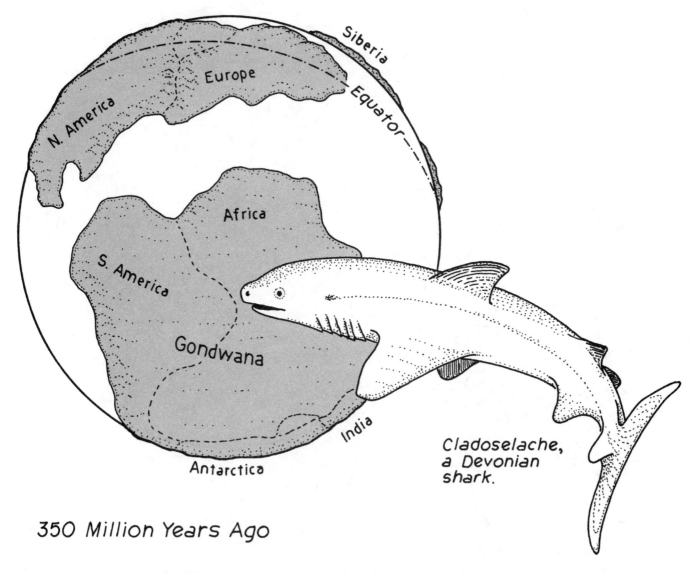

350 Million Years Ago

Cladoselache,
a Devonian
shark.

possibly since early in the planet's history. Continents have been arranged and rearranged, slammed together and rifted apart. Oceans have opened and shut, tipping their waters from basin to basin. New ocean floors have been extruded at the ridges and gobbled up at the trenches. No part of the present ocean floor anywhere on Earth is older than several hundred million years.

The map above shows the oceans and continents 350 million years ago, in the Devonian period. This is about as far back as we can go and reconstruct a map of the Earth's crust with any confidence.

It was during the Devonian period that plants consolidated their invasion of the land. Before this time, life existed only in the sea and the continents were gray and bare. The plants were followed onto the shore by the amphibians. The earliest insects—wingless species—

are found in Devonian rocks. But the Devonian truly belonged to the fishes, which enjoyed a time of great diversification. They came to dominate not only the oceans but also streams and lakes. Here in my drawing is *Cladoselache*, a Devonian shark that looks remarkably modern. *Cladoselache* is a common fossil in Devonian rocks on the south shore of Lake Erie, an area that was in those times a shallow continental sea.

Two large continents dominate the Devonian map. A unified North America, Greenland, and Europe lies astride the equator. This landmass had only recently been assembled by the collision of Europe and North America. The collision crunched up the Caledonian Mountains of Britain and Scandinavia. The climate in North America is tropical. In southern latitudes lies the supercontinent of Gondwana, embracing the future Africa, South America, Antarctica, India, and Australia. In northern seas, over our hori-

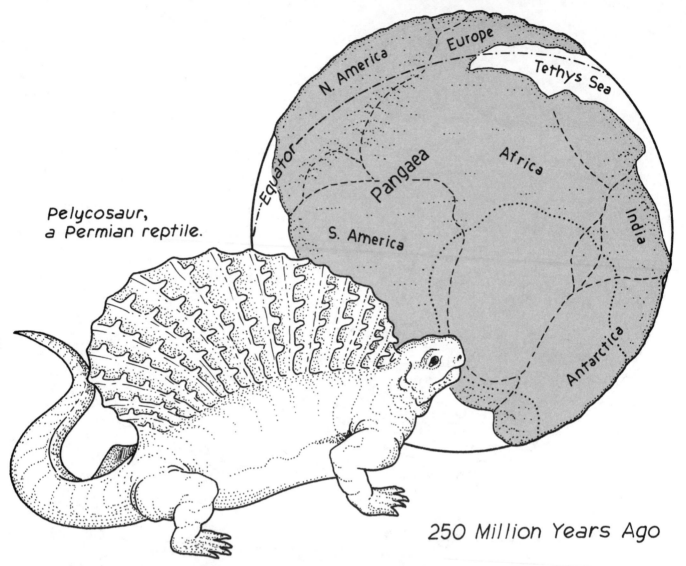

Pelycosaur,
a Permian reptile.

250 Million Years Ago

zon, Asia is being assembled by collision of smaller landmasses.

Now we jump ahead 100 million years to the Permian period. The map has changed dramatically. Africa (or rather the unified continent of Gondwana) has collided with North America in a supercrunch that pushed up the Appalachians, a range of towering snow-capped peaks. Along the top horizon of our map, Asia has smashed into Europe and raised the Ural Mountains. All of the major landmasses on Earth have been assembled into one supercontinent geologists call Pangaea ("all-earth"). The rest of the planet's surface is one great ocean, an arm of which—the Tethys Sea—intrudes deeply into the eastern flank of Pangaea.

The Permian coincided with one of the Earth's great ice ages. The dotted line on the map shows the limit of the continental ice cap that grew in Permian times near the southern pole. As the waters of the oceans were stored up in continental ice sheets, sea levels fell and waters retreated from continental interiors and margins.

The Permian saw the rise of the reptiles, ex-amphibians that had devised a way to reproduce without returning to the water. The trick, of course, was the egg, in which offspring could grow to essentially adult form without having to face the rigors of a dry world, cradled in a little bit of sea wrapped in a shell. The most spectacular of the Permian reptiles were the sail-backed pelycosaurs. The peculiar sail may have been a way for the reptile to regulate body temperature, a kind of radiator for shedding excess heat. The end of the Permian was marked by a wave of extinctions.

Another 100 million years brings us to the Jurassic period. I should perhaps have chosen a dinosaur to accompany the map of the Jurassic continents. The Jurassic was the time when the

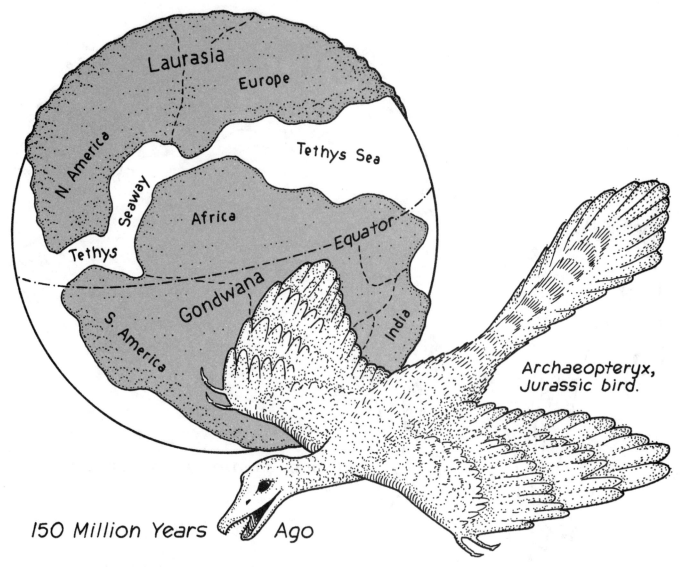

Laurasia
Europe
N. America
Tethys Sea
Seaway
Africa
Tethys
Equator
Gondwana
S. America
India

Archaeopteryx, Jurassic bird.

150 Million Years Ago

dinosaurs rose to prominence, a reign of supremacy that would last 150 million years. But I choose instead to illustrate the first feathered bird, *Archaeopteryx*, a creature that has taken to the air but resembles in obvious ways its reptilian ancestors. The birds would survive the later extinctions that obliterated the dinosaurs.

The same forces churning deep in the Earth that had previously assembled Pangaea now began to break the supercontinent apart. A rift opened near the line where an older Atlantic Ocean had been squeezed out of existence. As squeeze gave way to tension, the crust lifted and erosion of the Appalachian highlands accelerated. Along the rift where Pangaea broke apart, lava welled up from the mantle to form new sea floor. The present North Atlantic Ocean was born in fire as an extension of the Tethys Sea. A low-latitude seaway reached right around the globe, separating the northern supercontinent of Laurasia from a restored Gondwana.

During the Jurassic the central part of North America was submerged by a shallow sea. Further west, along the Pacific shore, intense geological activity began that would continue into our own times and ultimately create the Rocky Mountains and Sierra Nevada.

The Jurassic was a time of mild climate, a condition that seems to have suited the flourishing reptiles.

Now we jump again to 50 million years ago, and the map begins to look vaguely familiar. The Atlantic has continued to widen at the rate of an inch or so a year. In the north, rifting first separated Greenland from North America, and then more decisively wrenched Greenland away from Europe. North America has gone its own way, plowing up against the floor of the Pacific. The South Atlantic Ocean has opened as South America unzipped from Gondwana and drifted to the west. The Americas are not yet linked at Panama.

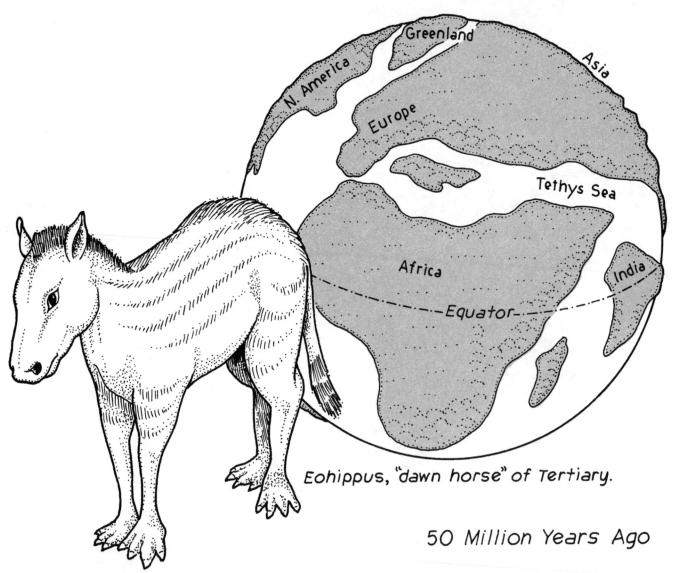

Eohippus, "dawn horse" of Tertiary.

50 Million Years Ago

Pangaea is in pieces. Gondwana has broken up and dispersed. Antarctica slips to the bottom of the globe, Australia drifts eastward, Madagascar slides south along the coast of Africa. One fragment of the old supercontinent, India, wanders northward across the equator on a collision course with Asia. The subsequent crash of continents will plow up the Himalayas.

Africa pivots on a hinge near Gibraltar and swings like a gate toward Europe, closing up the old Tethys Sea. The pieces of North Africa which will become Italy and the Balkans break off and race ahead to smash into the underside of Europe, throwing up the Alps. The famous Matterhorn is a little bit of Africa piled up on top of Switzerland. What is left of the Tethys Sea will become the Mediterranean.

On the continents, flowering plants and grasslands spread. Mammals replace the dinosaurs as the dominant animal. In response to the spreading grasslands, grazing mammals evolve and specialize. The first horses appear, represented here by dog-sized *Eohippus*.

With our last 100-million-year jump we leapfrog the present and take a look 50 million years into the future. The map is only an educated guess, based on present trends of plate motion.

The Mediterranean Sea has been squeezed out of existence, replaced by a range of high mountains. In Africa, the rifting which opened the Red Sea has continued northward. Along the Great Rift Valley of East Africa a sizable chunk of that continent has been wrenched away from its parent.

Further east, India has nudged yet more snugly against Asia, and Australia has plowed northward to affix itself as an extension of Asia. On the opposite side of the globe, southern California has glided northward along the San Andreas Fault—Los Angeles moving abreast of San Francisco—to become at last a Pacific island.

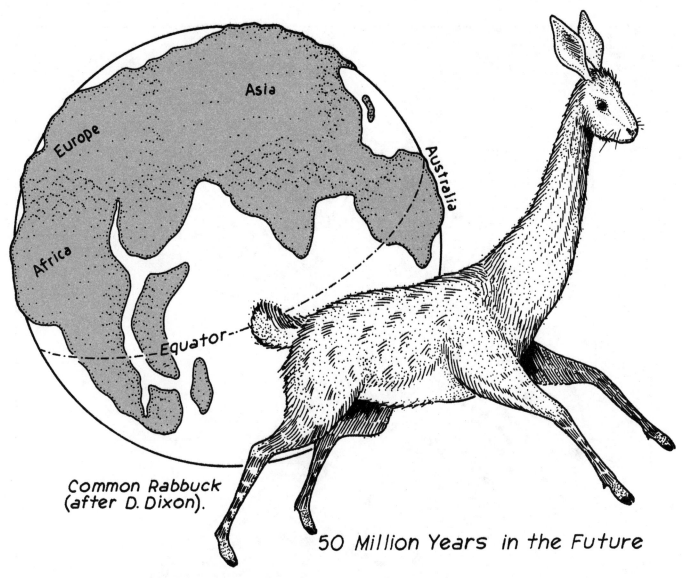

Common Rabbuck
(after D. Dixon).

50 Million Years in the Future

Still later, that sliver of California will become reattached to North America near Alaska.

It is more difficult to predict the future evolution of plants and animals or to make a reasonable guess for the fate of humans. I have represented future life on planet Earth with the rabbuck, a grazing descendent of the rabbit and a product of the fruitful imagination of Dougal Dixon. In his delightful book *After Man*, Dixon has created an entire zoo of the future by applying known principles of evolution.

No scientific discovery of recent times has been more exciting than the discovery of the dynamic nature of the Earth's crust. We caught a glimpse of the Earth's violent past at my outcrop of green volcanic rock slashed by pink granite. The full story of shifting plates has been pieced together from countless clues coded in the Earth's crust. In the next few chapters, we will explore some of the consequences for life of this geographical juggling of continents.

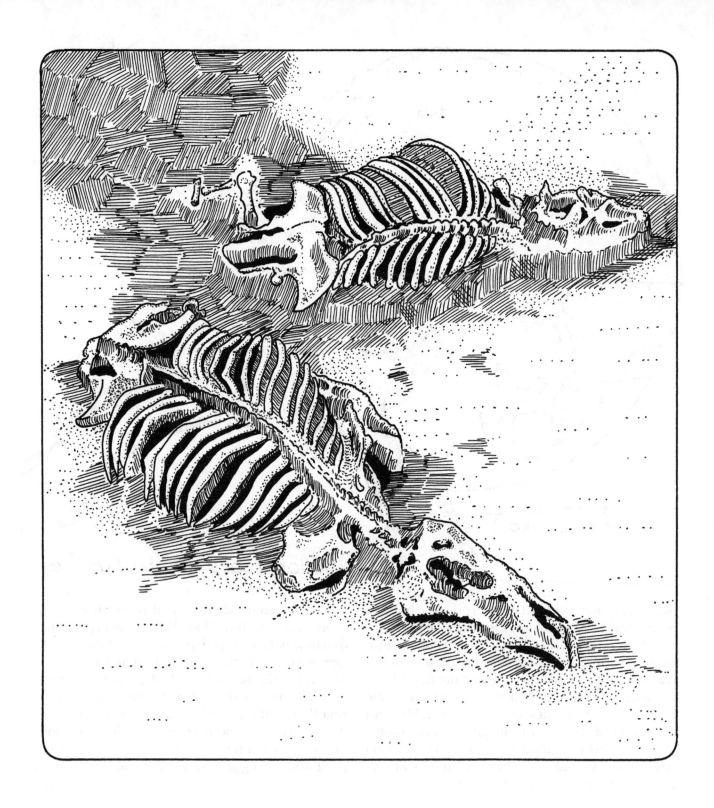

A Shroud
of Ash

Volcanic dust blasted into the upper atmosphere can modify climate on a global scale, and perhaps profoundly influence the history of life.

Some people prospect for gold. Others dig into the earth for coal or oil. Michael Voorhies, a paleontologist at the University of Nebraska State Museum, digs for fossils. He found his mother lode, struck his gusher, along the bank of a stream in eastern Nebraska.

The first hint of his treasure trove was the fossilized skull of a baby rhinoceros. The skull protruded from a bed of silver volcanic ash sandwiched between two layers of sandstone. Digging back into the hillside, Voorhies found the complete skeleton of the baby rhino. Later, assisted by his co-workers, he unearthed an entire zoo of extinct animals. The fossils were not of newly discovered species, but they were in a remarkably complete state of preservation.

The layer of ash that contained the fossils could be geologically dated to the Miocene epoch, 10 million years before the present. Ten million years ago Nebraska was a warm grassy savanna, not unlike parts of East Africa today. The animals Voorhies dug from the ash also seem more typical of Africa than North America. Rhinos dominated the dig. It is well known that great herds of rhinos roamed the western plains during Miocene times. There were also horses and camels. All of these animals later became extinct in North America. The rhino disappeared from the continent before the coming of hu-mans. The horse and the camel were pushed into extinction by the predations of the first human migrants to North America.

The animals exhumed by Voorhies were clustered about a waterhole when they were overcome by a suffocating ash fall. Here, in miniature, we can glimpse a mechanism for extinction that at other more violent times may have decimated plant and animal communities on a global scale.

The precise source of the ash cloud that snuffed out the lives of the creatures at the Nebraskan waterhole will never be known. The Miocene was a time of intense volcanic activity throughout the American West. The volcano that expelled the ash that choked the waterhole was certainly hundreds, perhaps a thousand miles away. All along the Pacific coast in Miocene times volcanoes were rising above rich subterranean reservoirs of lava. In the Columbia River Basin huge flows of a very liquid lava poured out of long fissures in the Earth's crust and spread out over hundreds of square miles of the surface. At the time of the calamity at the waterhole, the nearest active volcanoes to the Nebraskan site may have been in New Mexico.

The eruption that killed the Nebraskan rhinos, camels, and horses was on a scale far

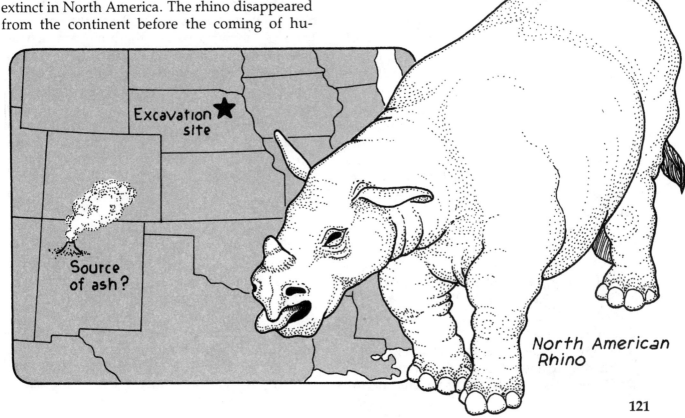

Excavation site

Source of ash?

North American Rhino

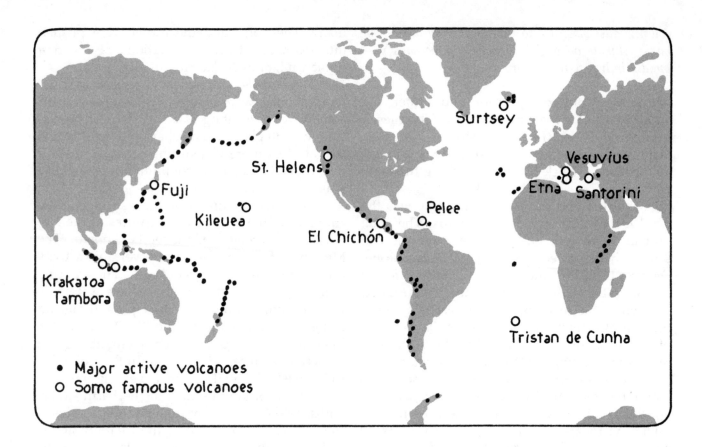

Major active volcanoes
Some famous volcanoes

Surtsey, Vesuvius, Etna, Santorini, St. Helens, Fuji, Kileuea, El Chichón, Pelee, Krakatoa, Tambora, Tristan de Cunha

greater than known eruptions of historic times. The May 18, 1980 eruption of Mount St. Helens dropped a layer of ash three inches thick 200 miles from the mountain. The ash that fell on the Nebraskan waterhole was ten times deeper from a more distant source. But the source of the heat that stokes Mount St. Helens is the same as the source that kept the Miocene volcanoes alive.

Along the western margin of North America, then as now, the continent grinds up against the floor of the Pacific in a battle of plates that has persisted for tens of millions of years. In the squeeze of plates, the Pacific sea floor is forced to slide beneath North America—a process known as subduction. Old slabs of Pacific sea floor even now lie beneath California, Oregon, and Washington. As they slide back into the mantle, these slabs of crust ignite the fires that have kept the American West volcanically alive. Friction, compression of the plates, changes in their mineral structure, and the transport of heat by convection may all contribute to the heat generated along the diving plates.

The rigid crust of the Earth is called the lithosphere. The lithosphere is typically about 50 miles thick—as thin, compared to the planet, as the skin of a grape or the shell of an egg. Below the lithosphere lies a seething cauldron of nearly molten rock, rock so hot that it churns and roils

in great slow loops of convective motion. This layer of the Earth's interior is called the asthenosphere. Pop the lid on this pressure cooker and the fire beneath our feet surges to the surface, flows out as sheets or streams of molten rock, or explodes violently into the air.

Most volcanic activity on Earth occurs at plate boundaries—at the cracks in the eggshell. The map shows a few of the hundreds of volcanoes currently considered active. Almost all of them lie on plate boundaries. Particularly obvious is the "Ring of Fire" around the perimeter of the Pacific Ocean. Of the volcanoes I have shown on the map, only Kileuea in the Hawaiian Islands lies far from a plate boundary. Kileuea stands above a "hot spot," a reservoir of heat in the upper mantle so powerful that molten rock (magma) pushes a plume of fire directly through the unbroken overlying crust.

Volcanic activity along plate boundaries has several expressions depending on the nature of the boundary. The cross section of the Earth I have sketched opposite, cutting across Mexico and the floor of the Pacific, illustrates two kinds of plate boundaries.

Along the rifted submarine ridge called the East Pacific Rise the plates move apart. The ridge is a typical divergent plate boundary. To the west of the ridge the huge Pacific Plate slides away on a two-hundred-million-year journey toward Asia. To the east, the tiny Cocos Plate makes a short quick trip to Mexico. A northern extension of this spreading axis has ripped away a sliver of the continent—Baja California—from the flank of Mexico.

The force that pulls the plates apart along the spreading axis has been hotly debated by geologists. One of the most popular theories assumes the existence of convection loops in the asthenosphere. At some places, along the ocean ridges, hot mantle material rises. In other places, at the subduction zones, cooler material sinks. As the plastic rock of the asthenosphere flows beneath the rigid crustal plates, it exerts a frictional drag that pulls the plates along.

As the plates move apart along the spreading axis, magma wells up from below to fill the rift. Pillows of lava ooze from the cracked Earth like toothpaste squeezed from a tube. The East Pacific Rise, a typical mid-ocean ridge, is one long continuously active volcano, patiently going about its business of building new ocean floor, new bony plate from the liquid marrow of the Earth. Only occasionally, as at Iceland in the Atlantic Ocean, does the hidden work of crust-making on the mid-ocean ridges explode above the surface of the sea.

The Cocos Plate is no sooner born than it finds itself pushed hard against the North American continent. With no place to go but down, the little slab of ocean floor is wedged back into the mantle, sliding to destruction along the Middle America Trench. As the plate descends it heats up, and it heats the lithosphere above it. Above the subducting plate molten rock floats toward the surface, melting its way through the brittle continental crust, finally erupting in the kind of explosive volcanic activity that has built up most of the famous volcanic peaks of the world. Mount St. Helens lies above a subducting plate, as do all of the active volcanoes of the American Northwest. Another wall of volcanoes stands along the coast of Central America, just inland from the Middle American Trench.

Among recent Central American eruptions was the April 4, 1982 explosion of the Mexican mountain known as El Chichón ("the lump-on-the-head"). The eruption took scientists by surprise. El Chichón stands well back from the sub-

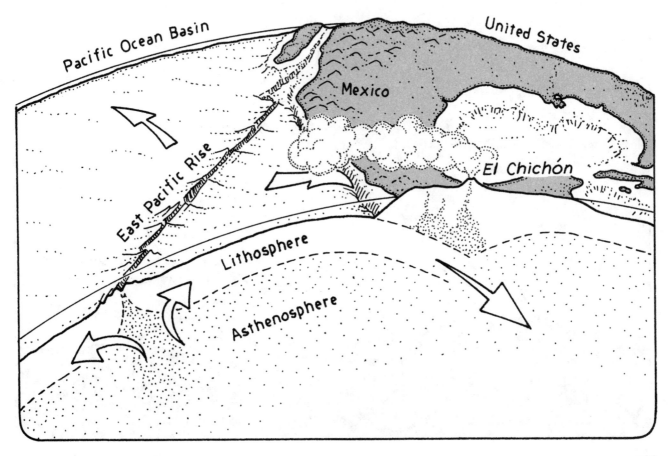

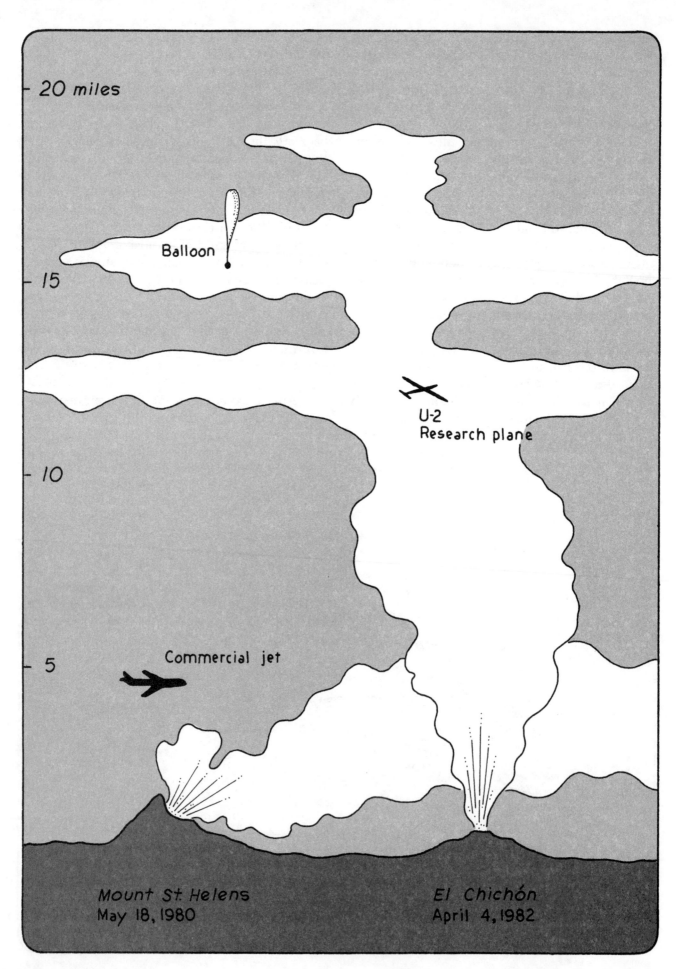

20 miles

Balloon

15

U-2
Research plane

10

5

Commercial jet

Mount St Helens
May 18, 1980

El Chichón
April 4, 1982

124

duction zone. The mountain's source of magma is something of a mystery, but the fires below the peak are almost certainly fed by the energy of the diving Cocos Plate.

El Chichón announced the impending catastrophe with rumblings, clouds of steam, and minor quakes. There were several preliminary eruptions. Local farmers took notice, but geologists were slow to react. The final cataclysm was like a cork popping from a bottle of champagne. A huge mass of ash and dust was blasted high into the air.

The volume of ash expelled by El Chichón was about equal to that expelled in the eruption of Mount St. Helens two years earlier. But the natures of the explosions were very different.

At Mount St. Helens, the "cork" of hardened lava in the throat of the volcano stayed put, and pent-up energy in the mountain blew out the side of the peak. A layer of ash was broadcast by the wind all the way to Montana, but not much dust went into the upper atmosphere. By contrast, the eruption at El Chichón went straight up, and fine powder was carried high into the stratosphere. High altitude circulation of the atmosphere carried the shroud of ash worldwide. The veil of volcanic dust was higher and more voluminous than any other volcanic cloud of this century.

The El Chichón dust veil became the most carefully studied cloud of all time. The cloud was sampled by airplane and balloon. It was probed by ground-based radar. It was examined by satellite. As I write, the cloud is still being sniffed and tasted by meteorologists armed with an impressive array of instruments.

The El Chichón ash dispersed into two stratospheric layers, one about 12 miles above the surface, another more massive cloud at 16 miles altitude. Winds carried the dust due west, circling the globe in 21 days. Within two months the upper cloud blanketed a band between the equator and latitude 30 degrees north, then spread into the southern hemisphere. The lower cloud had by then—like Sherwin-Williams paint—covered the globe.

Scientists studying the El Chichón cloud were curious to discover what would be its effect on the weather. A substantial warming of the stratosphere was soon observed. Some researchers believe stratospheric warming changed high altitude circulation patterns and contributed to the crazy weather experienced by inhabitants of North America during the winter of 1983–84. A one-half degree surface cooling in the northern hemisphere was predicted for 1984 and 1985. That doesn't sound like much, but such a small decrement in average temperature can significantly modify the Earth's weather. Predicting the precise effect of the El Chichón cloud at any given place, however, is probably an impossibility.

The El Chichón dust cloud was the greatest of this century, but it pales before the huge volume of pulverized rock spewed from the volcano Tambora on the East Indian island of Sumbawa in 1815. The Tambora eruption was the most powerful in recorded history. Almost 100,000 deaths were attributed to the eruption. The explosion lofted into the atmosphere 40 times more ash than Mount St. Helens and El Chichón combined. Thirty-six cubic miles of mountain literally disappeared into thin air. Most of the debris fell to earth in the East Indies, but the remainder stayed up for years, circling the globe, reddening sunsets, and reflecting sunlight back into space. Worldwide temperatures dropped for several years. In New England, 1816 was called "the year without a summer." It snowed in June. There were killing frosts in July and August.

The relationship between volcanic eruptions and average global temperatures is well established. One study of the advances and retreats of mountain glaciers over the past 100 years showed a close correlation with volcanic activity in the same latitudes as the glaciers. It is worth asking if the effect of volcanoes on climate could be severe enough to disrupt major plant and animal communities, perhaps causing major extinctions.

It has recently been suggested that the mineralogy of the iridium-rich clay layers at Gubbio and elsewhere, which some have interpreted as the result of an asteroid impact (see pages 106–111), are actually more consistent with volcanic activity. If this is so, then the extinction of the dinosaurs (including familiar triceratops, see next page) may have been caused by a long chapter of geological violence related to the clash of moving plates.

The engine of plate tectonics does not run smoothly. The geologic past has been punctuated by particularly violent episodes of rifting, mountain-building, and volcanic activ-

Triceratops

ity. There have been times when thin lavas have welled up from the ground along hundred-mile-long swarms of cracks and fissures, spreading across vast areas of the continents in broad flat sheets. The Columbia River Plateau of the American West and the Deccan Plateau of India were built up in this way. At other times, "hot spots" in the upper mantle have expelled huge volumes of lava onto the surface. Hawaii and Iceland are prominent examples of hot-spot activity. The late Cretaceous period, when the dinosaurs and many other species of plants and animals became extinct, was geologically a very "noisy" time.

In other words, the Earth is quite capable of "bursting at the seams," exposing the fiery cauldron that seethes beneath the planet's thin skin. The opaque drapery of gas, dust, and ash expelled in these episodes of violence could make the shroud of ash expelled by El Chichon seem like a wisp of gauze. During these times of unusual volcanic violence, the surface of the Earth would cool and darken. Photosynthesis could cease. Food chains could fail. Whole species of animals could starve and suffocate, falling in their tracks like the rhinos, camels, and horses at the Miocene waterhole in Nebraska, interred for the ages by the agent of their destruction.

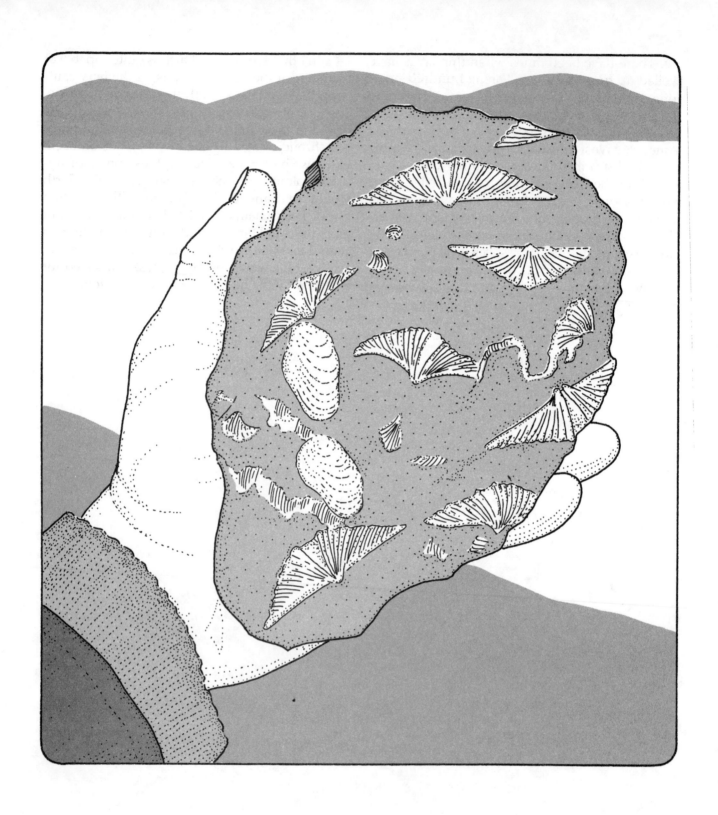

The Waters and
the Dry Land

Sea levels have changed relative to the land. Sometimes the waters have overlapped the continents, sometimes they have retreated to the deep ocean basins. In the process, shallow water environments for life have been created and destroyed.

My sister and brother-in-law live close by the Shokan Reservoir in the Catskill region of New York State. When the water level of the reservoir is low it is pleasant to walk along the exposed foreshore. It is also a good place to go looking for fossils.

The most beautiful of the fossils we have found are Devonian spiriferid brachiopods, extinct marine animals with a hinged two-part shell like an oyster or a clam. The calcareous shells of the spiriferids have left a snowy-white residue in the dark grey sandstone in which the fossils are found. Against a slab of grey stone, such as the one I have drawn at left, the spiriferid fossils have the aspect of a flight of tiny white birds.

The sedimentary deposits around the Shokan Reservoir are 350 million years old, dating from the Devonian Period of geologic time. Spiriferids appeared in the Ordovician, flourished in the Devonian, and persisted until the Jurassic Period before disappearing from the fossil record. Altogether, the group endured for 300 million years.

But what are these graceful sea creatures doing in the Catskill Mountains, 80 miles from the present seashore? Even in Devonian times the region of the Shokan Reservoir was far from a continental margin. The answer lies in the way sea levels have changed relative to the continental platforms.

If one ignores continental mountain ranges and the deep ocean trenches, there are two basic levels to the Earth's crust—the continental platforms and the floors of the oceans. The two levels differ in elevation by about 3 miles. The continental crust of the Earth and the rocks of the ocean floors float in a kind of equilibrium on the hot plastic rocks of the asthenosphere. The continents float higher for the same reason a block of balsa wood floats higher in water than a piece of oak. The rocks of the continental crust are less dense than the rocks of the ocean floors.

It so happens that there is just enough liquid water on the planet to fill the ocean basins to the brim. A relatively small rise in sea level can cause the oceans to overflood the continents, creating the shallow waters of the continental shelves and of inland continental seas such as Canada's Hudson Bay or Europe's North Sea. The encroachment of the sea onto the land is called transgression. The retreat of the shoreline toward the ocean basins is called regression.

Before we consider why the ocean levels rise and fall, let's look at changes that have occurred in the past. On the following three pages I have sketched areas of dry land and shallow sea on the continental platform of North America at three times in the past. The outline of the present continent is added for reference. Bear in mind that maps such as these rely on the interpretation of complex geological data and should be considered schematic only.

The first map shows North America at the time when our Devonian spiriferids flourished in the region of the present Catskill Mountains. The Devonian was a time of generally high sea levels and most of the North American continent was submerged by shallow waters.

The eastern margin of the Devonian continent lay along a convergent plate boundary. The floor of an older "Atlantic Ocean" was being forced beneath North America as the continent of Africa approached from the east. There may have been a line of volcanic islands offshore, as there is today where ocean floor is being consumed along the western margin of the Pacific.

The pressure of the converging plates raised a highland rim on the eastern edge of the continent. Erosional debris washed down from the highlands into the shallow water-filled basin directly to the west. These sediments would become the sandstones and mudstones that enclose our spiriferid fossils. In the subsequent crunch of continents—Africa colliding with North America about 300 million years ago—these sediments and the fossils they contain would be lifted into the ancestral Catskill Mountains.

There was another highland rim near the western edge of the continent. Generally, however, the Devonian was a geologically quiet time in North America. The continental interior was a platform of low relief, rimmed with highlands on east and west like a shallow water-filled dish. Here and there islands lifted above the waters. The islands may have provided laboratories for life's early experiments out of the water. Plants crept onto the sandy shores, and amphibians and insects followed closely.

But in spite of life's early successes on land, the Devonian period truly belonged to the fishes. The fishes consolidated their enduring tenure in the Earth's oceans, lakes, and streams. I have represented the early fishes with a member of the jawed and armored placoderms. Some placoderms became large and voracious predators, reaching lengths as great as 30 feet. The species illustrated here is *Bothriolepis*, a proficent gleaner of food from the floors of rivers and lakes.

The second map takes us ahead 200 million years to the Jurassic Period. A long, eventful chapter of geological violence in the eastern part of the continent has passed in the interval. The Appalachians have been thrown up by the collision of Africa and North America. The convergence of continents has given way to tension. Rifting east of the line of collision has begun to open up the modern Atlantic Ocean. The Atlantic opened from south to north like a zipper, and

in Jurassic times Europe and Greenland were still solidly attached to North America.

The focus of geologic activity in North America has shifted to the west, where a new battle of plates is under way. The continent of North America pushes against the floor of the Pacific, which dives to destruction along an offshore trench. Inland, a line of volcanoes attest to the subterranean violence.

The Jurassic continent even more resembles a dish, with a pronounced mountain rim on east and west. Global sea levels have dropped, but much of the continental interior remains submerged. Where the Rockies stand today, a great embayment reaches down from the Arctic Ocean and another pushes up from the Gulf of Mexico. The northern embayment is called the Sundance Sea.

Along the swampy margins of these inland seas the dinosaurs flourished. The thunder-footed sauropods—including the familiar brontosaurus—thrived in the warm, shallow-

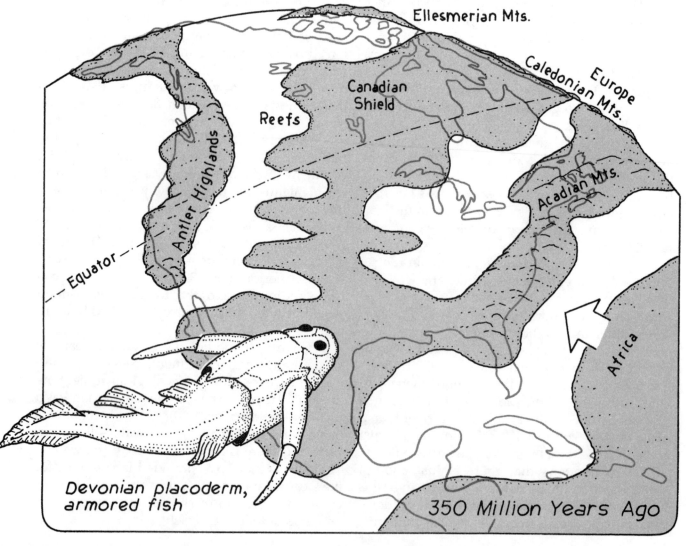

Devonian placoderm, armored fish

350 Million Years Ago

water environment. Later, when the Rockies were pushed up from the floor of the Sundance Sea, the bones of dinosaurs were lifted to mountain tops.

The plesiosaurs, reptilian cousins of the dinosaurs, took to the waters in Jurassic times and were soon distributed worldwide. Limbs of the great beasts were transformed into paddles. Slender curved teeth were adapted for snaring fish. These agile, long-necked swimmers were undisputed masters of the Jurassic seas.

During the late Cretaceous Period, rising seas advanced vigorously onto the continent, cutting North America from north to south with a continuous inland waterway. Sea levels stood highest on the continents about 80 million years ago. Since that time the seas have been in full retreat. The third map shows the approximate shorelines of North America 50 million years ago, during the Eocene Period of the Cenozoic Era.

The western margin of the Eocene continent is volcanically active. The Rocky Mountains are born as the squeeze of plates along the western coast slowly crumples the continent eastward. There are basins between the newly uplifted highlands, some filled with lakes, that are dumping grounds for sediments eroded from the mountains. The most economically important of these deposits are the oil shales, muddy sediments rich in the remains of organic life that flourished in the basin lakes.

In the east the geological drama is past, and the Appalachians continue their slow decay. The waters of the widening Atlantic Ocean overlap the continental shelf more deeply than presently. In those shallow waters sediments are deposited that will become the coastal plain of the southeastern United States.

It is worth noting that even as late as the Eocene several prominent features of the modern map of North America are yet to come into existence. The Great Lakes and Hudson Bay are

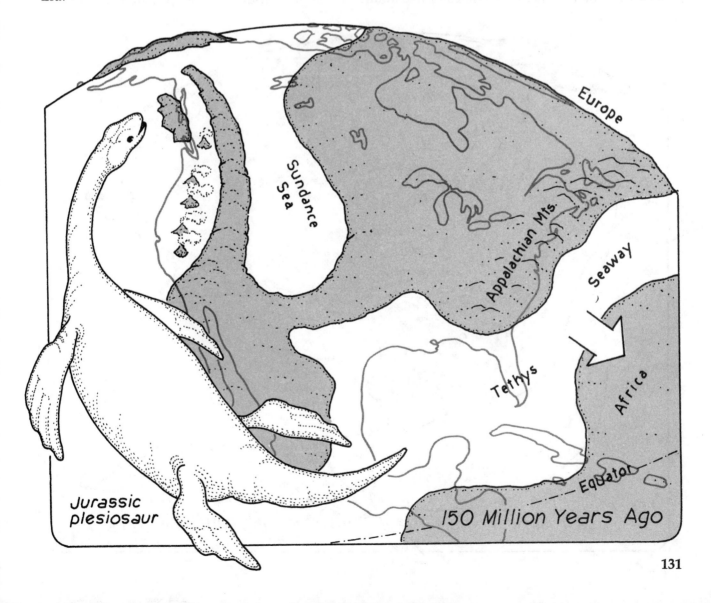

Jurassic plesiosaur

150 Million Years Ago

conspicuously absent. Those great bodies of water are artifacts of the recent ice age (see Chapter 20); they are no more than 10 thousand years old. You will also note that fifty million years ago Baja California had not yet been sliced from the side of Mexico, nor were North and South America connected at Panama.

By Eocene times some mammals had managed to adapt to life in the water. The whales appear quite suddenly at about this time, as fully developed ocean-goers. The whales replaced the extinct dinosaurs as the largest animals on Earth. *Basilosaurus*, illustrated here, reached lengths of 60 feet.

The three maps on these pages are like random snapshots of relative levels of land and sea. A time-lapsed motion picture of the continent over the past 600 million years would provide a more dramatic record of shifting shorelines. Sometimes we would watch the oceans lapping higher on the continents. At times of maximum

transgression, mountains would seem to stand as islands in a global sea. At other times the waters would draw back, exposing broad areas of the continental shelves. A rough curve of past sea levels is shown opposite, with present-day shorelines as a reference. It is clear that compared to most of the past, we live today in an era of dry land.

There has been a lively controversy among earth scientists regarding the cause or causes of marine transgressions and regressions, and whether or not there has been a long-term trend toward lower sea levels.

A change in relative sea level can result from a change in the amount of water in the ocean basins, or from a change in the volume of the basins, or from a change in the elevation of the continents.

The amount of water in the ocean basins is an inverse function of how much water is stored on the continents in the form of ice. For instance, if the present-day ice cap on Antarctica melted,

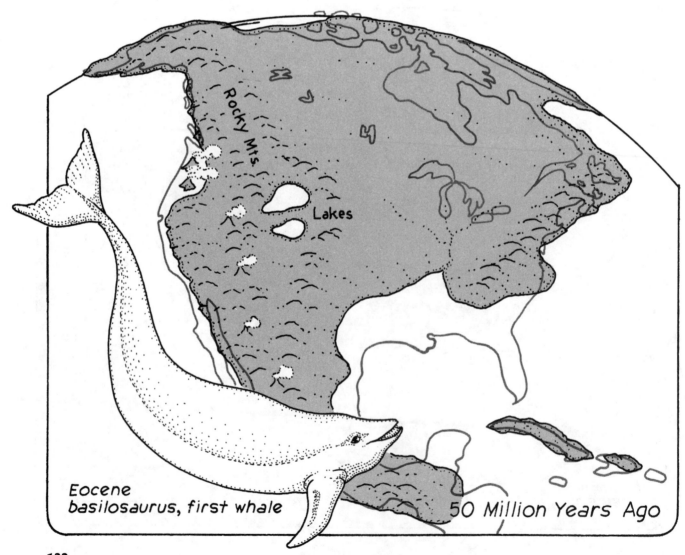

Eocene
basilosaurus, first whale

Rocky Mts.

Lakes

50 Million Years Ago

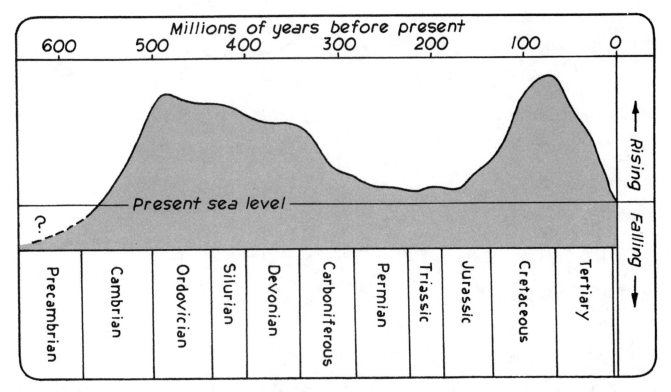

Millions of years before present

600   500   400   300   200   100   0

Present sea level

Rising
Falling

?

Precambrian | Cambrian | Ordovician | Silurian | Devonian | Carboniferous | Permian | Triassic | Jurassic | Cretaceous | Tertiary

the waters of New York harbor would rise to Lady Liberty's chin. New York, Rio, Tokyo, and other great coastal cities would be submerged. Add the water stored in the Greenland ice cap and the inundation would be compounded. Alternatively, when great ice sheets lay on the northern continents twenty thousand years ago, the seas retreated to the continental slopes and the site of New York City was hundreds of miles from the ocean shore.

The weight of a mantle of ice on a continent will depress the elevation of the continent, just as a raft floats lower in the water when a passenger steps aboard. The last of the great continental glaciers on North America retreated only 10 thousand years ago. The continent has not yet fully rebounded to its preglacial elevation. Hudson Bay, for example, stands where the continent was depressed by ice. As the continent rebounds, the waters of the Bay will be tipped back into the ocean.

Ice is not the only control on sea levels. Plate tectonics is always at work flexing the continents, crumpling their margins, rearranging their bulk (perhaps adding to their bulk), and

otherwise modifying their posture with respect to the sea. It is widely acknowledged that periods of rapid plate divergence at ocean ridges correspond to times of substantial transgressions of the sea onto the land. The high broad ridges that rise above sea-floor spreading centers reduce the volume of the ocean basins and cause the waters to lap higher on the land. There is evidence that the very high sea levels of the Late Cretaceous period (see graph) correspond to a time of intense volcanic activity along subduction zones. Activity at subduction zones presumably accompanies vigorous sea-floor spreading at mid-ocean ridges and decreased volume in the ocean basins.

Whatever the cause of transgressions and regressions of the sea, the effect on life has been traumatic. With the rise and fall of the sea, shallow-water habitats for life are created and destroyed. Creatures too narrowly adapted to a particular ecological niche are placed in mortal danger. Changing sea levels are one more way the engine of plate tectonics keeps changing the rules by which plants and animals play the game of survival.

The Great American
Interchange

The game of life is played in a changing environment. The creation of an isthmus at Panama is one example of how plate tectonics can shuffle the deck.

A few days ago we found a possum trapped in a window well of the science building at my college. We helped the animal make good its escape, mindful of the gaping jaws and fierce set of teeth. Freed from its temporary prison, the possum scampered into the brush without so much as a "thank you."

New Englanders have seen a lot of possums lately, especially as casualties on the road. The possum (or more formally, opossum) arrived in Massachusetts 17 years ago and has been increasing in numbers ever since. The animal was nudged into our region as shopping malls and suburban development reduced its more southerly habitats. Interstate 95, which runs along the east coast, provided a ready avenue of migration.

The possum is not an attractive domestic scavenger. It has the scruffy, unkempt look of a Bowery bum and the reputation, probably deserved, of being stupid. Certainly the possum lacks the cuddlesome charm of its partner in backyard crime, the raccoon.

But it would be hard to fault the possum for its resourcefulness. The rat-tailed marsupial has sometimes been called a "living fossil." It has survived with little change for tens of millions of years, since the time of the earliest mammals. The first possums shared the Earth with dinosaurs.

Possums, like other marsupials, are born in a very immature state and must spend a nursery residence in the mother's pouch. Placental mammals, on the other hand, come into the world with "all systems go." Marsupials have not done well in competition with the more intelligent and adaptable placentals. Generally, marsupials have flourished only in geographical isolation. The wombats, bandicoots, koalas, and kangaroos of Australia are obvious examples of marsupials which have done well in the absence of stiff mammalian competition.

The possum is the exception to the rule. The possum has gone from success to success, often in competition with placentals. The story of this little South American native's long trek to

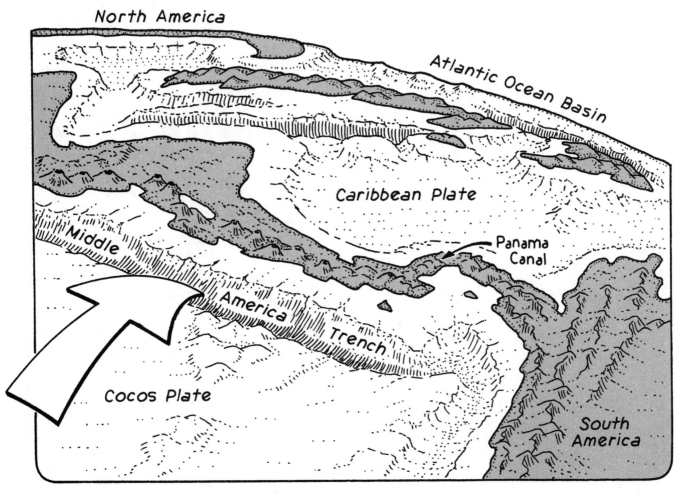

New England is a tale of staying power and pluck. The story begins millions of years ago when the Panama Isthmus began to rise from the floor of the sea.

The crust of the Earth is fragile—eggshell thin, really—and is torn, twisted, and dragged about by powerful forces stirred up by the planet's internal heat. During the Earth's long history the continents have been arranged and rearranged, slammed together and rifted apart. Oceans have opened and shut, tipping their waters from basin to basin. Every few hundred million years the surface of the globe has been completely revamped. The familiar map on the schoolroom wall is only the most recent snapshot of the tumultuous history of our planet. Life must adjust its drama to this shifting stage.

When the first mammals appeared on Earth, the continents were united in a single landmass geologists call Pangaea ("all-earth"). Not long after the mammals made their debut, Pangaea broke apart and the continents began their drift to their present positions. For a hundred million years South America was on its own, unconnected to any other landmass. The isolated continent evolved distinctive communities of plants and animals. The mammals of South America, like the mammals of Australia today, included many marsupials.

Beginning about twenty million years ago, the part of the Earth's crust between North and South America was caught in a geological squeeze. The tiny Cocos Plate was pushed into the mantle below the Caribbean Plate. Above the subducting plate the sea floor was crumpled upward and volcanoes injected lavas into the crust. By three million years ago a solid link had been established between the two continents.

Going South

Skunk

Llama

Tapir

Traffic across the new land bridge moved both ways. Among the earliest migrants to the southern continent were raccoons, rats, and mice, waifs that somehow rafted from island to island even as the link was rising. When the bridge was in place, skunks and peccaries walked south. Then came dogs, wolves, foxes, cats, bears, llamas, deer, horses, camels, tapirs, and the elephantlike mastodons. The llama and the tapir are now native only to the continent that became their new home. The mastodon is gone forever. Horses and camels evolved in North America, walked south across the bridge, and then became extinct on both continents.

The northern invaders were spectacularly successful in the south. Native marsupials fared poorly in competition with placentals from the north. All but two species of southern marsupials were wiped out.

A changing physical environment in South America may have aided the northern invaders. The geological squeeze that crumpled up the Panama isthmus affected the entire western margin of the Americas, creating a mountain wall from Alaska to Tierra del Fuego. During the interval when the Panama bridge was rising, the Andes Mountains of western South America doubled in elevation. The towering range acted as a barrier to warm, moisture-laden winds from the Pacific, deflecting them upward. As the air cooled, the moisture it carried was precipitated as rain or snow on the western slopes of the mountains. It was air wrung dry that crossed the peaks, creating a "rain shadow" to the east of the range. Habitats in southern South America changed from grasslands and woodlands to drier forests and deserts or semi-deserts. The northern immigrants were better able to in-

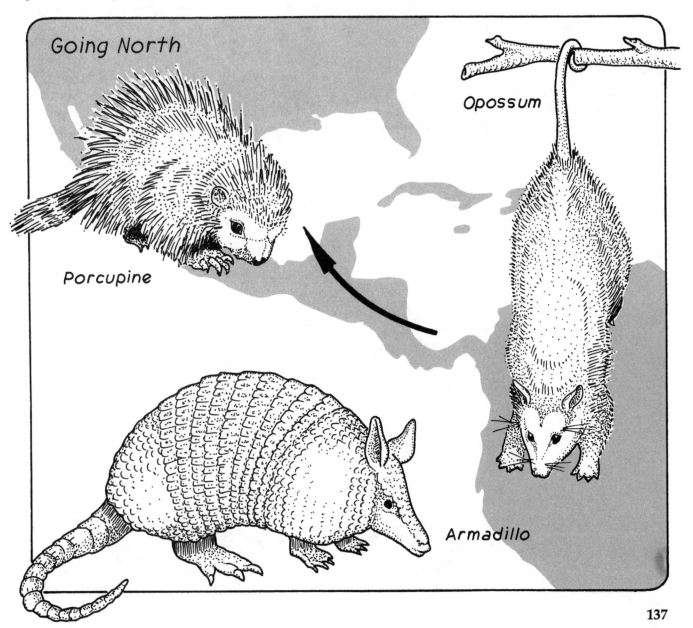

Going North

Porcupine

Opossum

Armadillo

sinuate themselves into the new ecological niches than were the southern natives.

Meanwhile, South American animals found the narrow mouth of the connecting funnel and made their way north. Armadillos blazed the trail along with porcupines and capybaras. In the next wave of immigrants were the opossums and ground sloths. Later arrivals included marmosets and monkeys, tree sloths and anteaters.

The southerners that trekked across Panama faced troubles in the north. Few species managed to establish themselves permanently on the new continent. The giant ground sloth from South America would seem to have had every prospect for success. It was the size of a small elephant and as powerful as a grizzly. Another promising migrant was the glyptodon, sometimes called the "mammalian tank." The glyptodon carried heavy armor and wielded a thick tail spiked at the end like a medieval mace. But the size of the ground sloth and the armor of the glyptodon were insufficient to protect these formidable creatures from perils they met in the new land.

Today, North and South America share many families of animals, most of them originally from the north. The convergence of families was brought about by traffic across the land link at Panama. In the adjacent seas, an opposite effect prevailed. For marine animals, the land link was a dam, a barrier to migration. At about the same time that continental families began the trek toward convergence, marine species on opposite sides of the isthmus began to evolve along divergent paths. The trend is most evident in the fossils of single-celled marine organisms called forminifera. The fossils can be retrieved by drilling into sea-floor sediments. The story of isolation and divergence is also evident even in the seashells you might pick up on the Caribbean and Pacific shores of Panama.

Some scientists believe the creation of the isthmus at Panama was the trigger for a series of ice ages on the northern continents. Before the completion of the link, a part of the North Atlantic Equatorial Current passed through the Caribbean into the Pacific. When the isthmus was in place, this warm-water current was deflected northward, reinforcing the Gulf Stream. Two conditions are necessary for an ice age to begin—cool continental temperatures and sufficient moisture in the air to fall as snow. A strengthened Gulf Stream may have contributed to the second requirement and helped to initiate glaciation in northern latitudes. Southern species that moved north across the land bridge were unprepared for the rigors of an ice age.

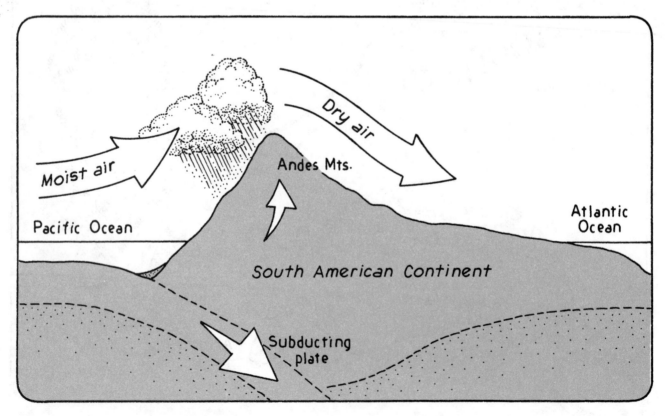

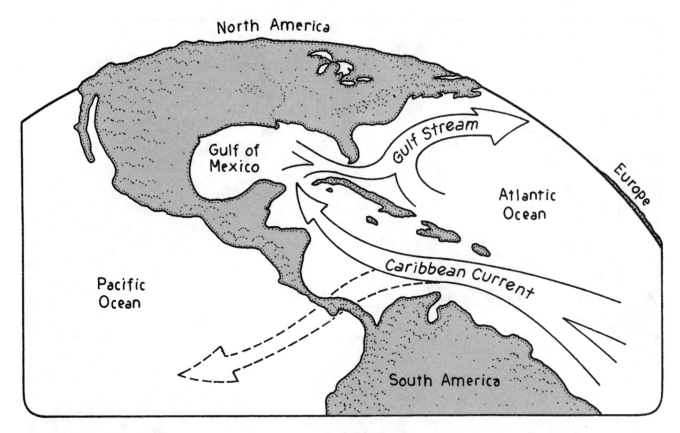

The creation of great ice sheets on the northern continents posed yet another precarious situation for life. The new threat challenged northern and southern species alike. That new threat will be the subject of the next chapter.

Of the southern mammals that moved into North America across the Panama land bridge, only the armadillo and the possum survive today. And only the possum preserves the marsupial line in the north. Mountains rose and fell. Ocean currents were deflected. Continents were piled with ice. Throughout the geological turmoil the plucky possum prospered. It competed successfully with northern placentals. It adapted to changing environments and climates. It outsmarted and outlasted human predators. It continues to expand its range and has brought at last to New Englanders an exotic glimpse of other continents and other geological eras.

We have seen, lately, many opossums dead in the road. But the little marsupial will no doubt outlast the automobile as well.

A World
of Ice

The Earth's climate is milder today than in the recent past. But by comparison with most of Earth history, we are living in an "ice age."

You will recall a visit we made in an earlier chapter to an outcrop of bedrock near my home in New England. There we saw evidence of hundreds of millions of years of geological upheaval, the clash of continents, the building of mountains, and eons of erosion. Let's go now to another outcrop near my home, the wooded rocky knoll I have sketched at left. Here we find evidence of a different kind of geological violence.

At the crest of the knoll there is a broad expanse of exposed bedrock, true crust of the Earth. The rock is a dark, fine-grained volcanic, the same green rock we found at the earlier site. Patches of the exposed surface are as flat and smooth as if they had been abraded by an electric sander. Close examination—especially when the rock is wet—reveals fine scratches (striations) on the flat surfaces of the rock. All of the scratches run in a north-south direction.

The striations were made by moving ice! And apparently they were made not long ago, geologically speaking, for the fine scratches have not yet weathered away.

The scratches are not the only evidence in this place of moving ice. Perched on the outcrop of green volcanic rock are huge glistening boulders of granite, some weighing as much as 20 tons. The boulders are of a kind of rock found five miles to the north of the wooded knoll. Geologists call these boulders "erratics," meaning they are not at their place of origin. They were broken from granite outcrops and carried

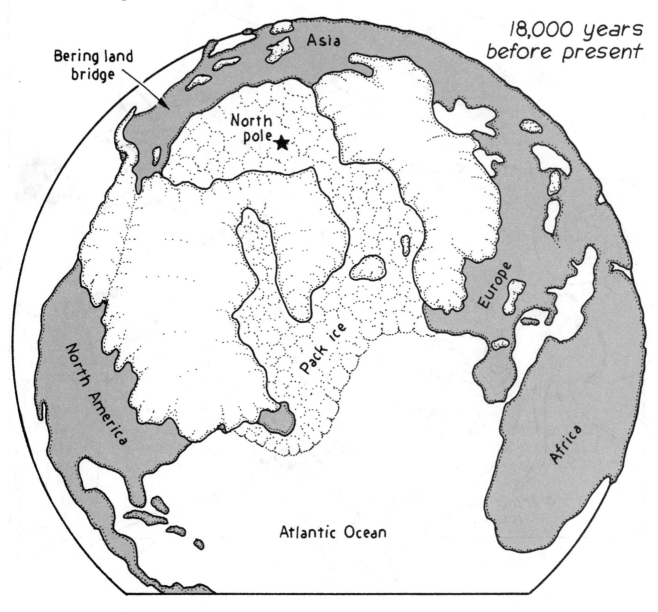

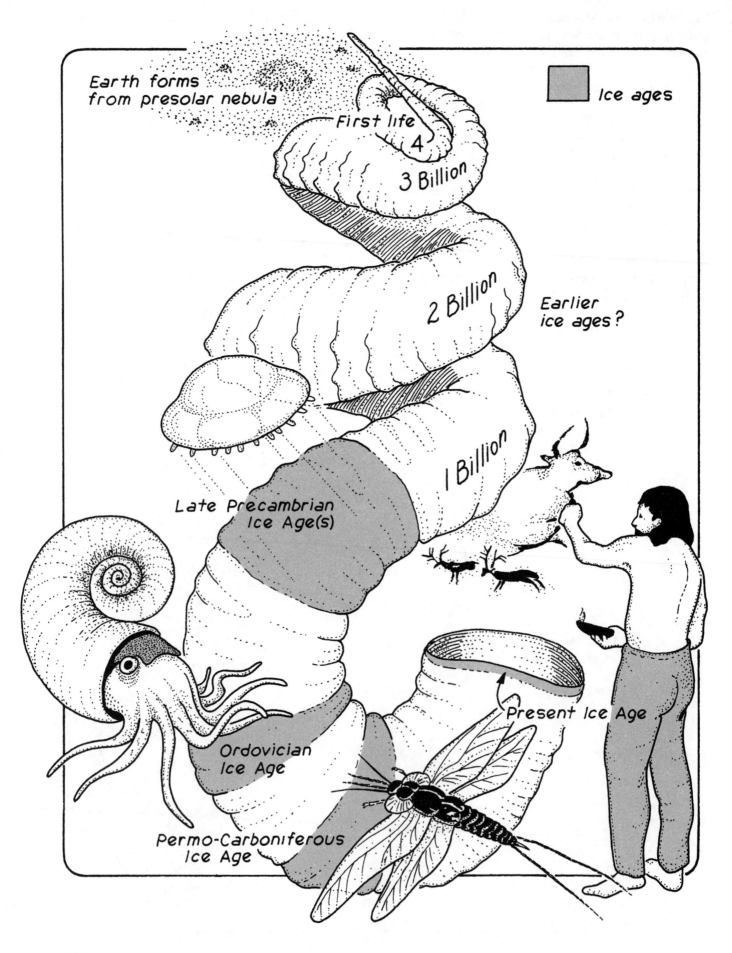

Earth forms from presolar nebula

Ice ages

First life
4
3 Billion

2 Billion

Earlier ice ages?

1 Billion

Late Precambrian Ice Age(s)

Ordovician Ice Age

Present Ice Age

Permo-Carboniferous Ice Age

south by the ice. When the ice melted, the boulders were dropped where we find them today.

Putting together clues such as these, geologists have reconstructed the great ice ages of the past.

We are apparently living in an ice age now. For most of the past 3 million years, large parts of Canada and the northern United States have carried a mantle of ice more than a half mile thick. As recently as 12,000 years ago New England still felt the presence of the glaciers. The present ice-free conditions may be only a brief respite in a continuing regime of ice.

Two things are necessary for a glacier to grow. There must be sufficient moisture in the air to insure plentiful winter snowfalls, and the climate must be cool enough so that not all the snow that falls in the winter melts in the summer. Year by year the depth of snow builds up. Eventually the weight of the accumulation turns the snow into ice. When the ice can no longer support its own weight, it squeezes out at the margins, like a slowly flattening blob of dough or putty. The ice keeps moving outward as long as the conditions are met for continued accumulation.

For most of the past 3 million years these conditions have been satisfied at places in northern Canada, Scandinavia, and Siberia. The ice crept outward from these centers of accumulation like bulldozers, merging into continent-spanning sheets, deepening and rounding valleys, ripping boulders from hillsides, pushing over forests, grinding and polishing the surface of the Earth. No animal, no plant, could resist the relentless push of the ice.

When the ice reached the open sea it pushed out onto the water and broke apart into bergs. On the continents the ice pushed south until it nosed its way into a climate where the rate of melting kept pace with the inflow of ice from the northern centers of accumulation. In North America the line of farthest advance was near the course of the present Missouri and Ohio Rivers. In New England the terminus of the ice was along the line of Long Island and the offshore islands of Massachusetts.

There have been many advances and retreats of the northern glaciers during the past 3 million years. That entire long period is sometimes called "The Great Ice Age," to distinguish it from the recurring pulses of glaciation. We are apparently still in the Great Ice Age, although in something of an intermission.

There have been at least four "great ice ages" in the history of the Earth, each punctuated with recurring episodes of continental glaciation. The great ice ages seem to occur at intervals of several hundred million years and last for tens of millions of years.

The farther back in time we go, the more difficult it is to reconstruct details of earlier great ice ages. The reading of clues is complicated by the movement of continents and the burial or erosion of rocks that might show evidence of the ice. The sketch below shows the sort of clues geologists look for. Scratches and grooves on the upper surfaces of rocks that were exposed at the time of the glaciation are certain evidence of moving ice. These scratched surfaces are often covered and protected by a rock called tillite, formed from the consolidation of glacial debris.

Once again I have used my mollusc time line to indicate the great ice ages of the past. We have recently entered a time of cooler climate and spreading ice. Seventy million years ago, when dinosaurs ruled the Earth, the global cli-

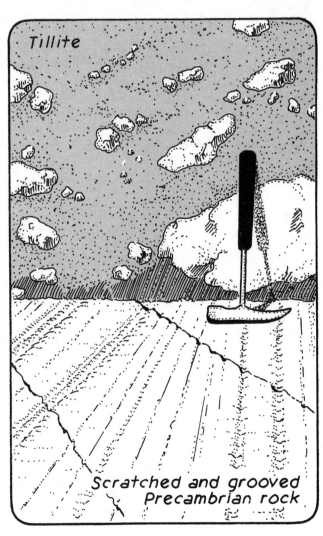

Tillite

Scratched and grooved Precambrian rock

mate was significantly warmer than at present. The climate in Labrador, for instance, was not unlike the climate today in South Carolina. Antarctica and Greenland were ice free. The Sahara was a fertile grassland. Then, at about the time of the dinosaur extinctions, global temperatures began a slow decline. By 14 million years ago, and possibly much earlier, Antarctica carried an ice cap. By 10 million years ago mountainous regions of the northern hemisphere were glaciated. About 3 million years ago conditions were right for huge glaciers to build up on the northern continents. Since that time, the ice has advanced and retreated in great pulses of activity. At times the glaciers reached as far south as Kentucky. At other times they retreated to permanent redoubts in Greenland and Baffin Island. We are presently living through a period of glacial retreat, a brief intermission (perhaps!) in a long, long age of ice. Only 10 thousand years ago our Cro-Magnon ancestors sheltered from the cold in caves of Spain and southern France and hunted on the frozen tundra at the margins of the ice.

The ice began to draw back about 18 thousand years ago and 10 thousand years ago it was in full retreat. The lingering coolness of the most recent ice age lasted until the eve of recorded history.

Few subjects have sparked more lively controversy among earth scientists than the cause or causes of the ice ages. The theories are almost as numerous as the scientists who speculate. Only one thing is indisputable: the Earth's weather machine is enormously complex, and global climate depends upon a daunting tangle of circumstance.

In an earlier chapter we discussed the possibility that the Sun might not be a constant star. Variations in the energy output of the Sun might indeed be a factor in the Earth's climate. But most scientists look closer to home for an explanation of the ice ages. Surely, the dynamic nature of the Earth's crust can be cause enough for variations of climate.

What geological events might have instigated the most recent Great Ice Age, the one that began 3 million years ago and continues today? There were many crustal changes during the past 50 million years that might have led inexorably to a drop in global temperatures and the onset of an age of ice.

In the south, Antarctica was drifting toward the bottom of the globe. The presence of a large landmass at the southern pole insured glaciation of that continent. Eventually, 5 million square miles of Antarctica were crusted with ice. Ice reflects more of the Sun's incoming radiation than does land or water, turning back to space potentially warming rays of sunlight. When Ant-

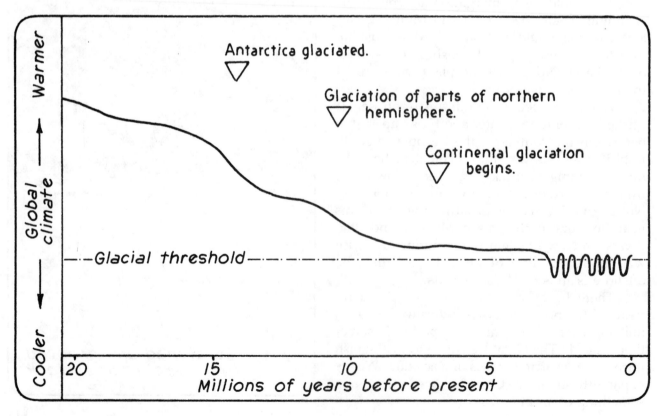

arctica became glaciated, the loss of solar radiation precipitated a drop in global temperatures.

At some point in its southern drift, Antarctica broke a land link with South America and instigated new patterns of circulation for currents in the southern oceans.

Meanwhile, the northern continents were also drifting toward cooler polar latitudes. In addition, rifting was slicing up between Greenland and North America and between Greenland and Europe, opening the Arctic Basin to currents from the North Atlantic. Uplift at the margins of the rifts changed the topography of these northern lands from low plains to elevated plateaus with mountainous rims. Large areas of land were lifted into cooler temperatures.

Three million years ago the completion of the isthmus at Panama deflected tropical ocean currents northward, bringing moist air to northern latitudes. Now both conditions for glaciation were met: cooler continental temperatures and plentiful winter snows. Ice began to accumulate. The closing of the seaway at Panama may have been the last link in a chain of events that led the world inevitably to the brink of massive continental glaciation.

The glacial threshold was passed about 3 million years ago. Since that time global climate has oscillated about the threshold and the ice has alternately advanced and retreated on the continents. In recent years it has become almost certain that the "pacemaker" of these glacial oscillations is the changing orbital relation of the Earth to the Sun.

The Earth's posture with respect to the Sun changes in several ways, illustrated schematically on the drawing below. First, the eccentricity ("off-centeredness") of the Earth's orbit varies in a cycle lasting 100,000 years. Second, the tilt of the Earth's axis wiggles up and down about 3 degrees in a 43,000 year cycle. Finally, the axis of the Earth wobbles as the Earth spins, like the axis of a top, in a cycle lasting 23,000 years. These variations work together in a complex way to influence the distribution of the Sun's radiation over the face of the globe. Although the net effect is small, it seems to be enough to cause continental glaciers to advance or retreat—if the planet is ready at the glacial threshold!

There are feedback mechanisms that might amplify Sun-induced glacial cycles. For example, spreading ice increases the reflectivity of the planet, turning sunlight back to space and accelerating cooling temperatures. Or, in the other direction, a decrease in ice volume leads to higher sea levels, which enhance the attack of the sea on land ice and a further decrease in ice volume. Many other kinds of feedback loops

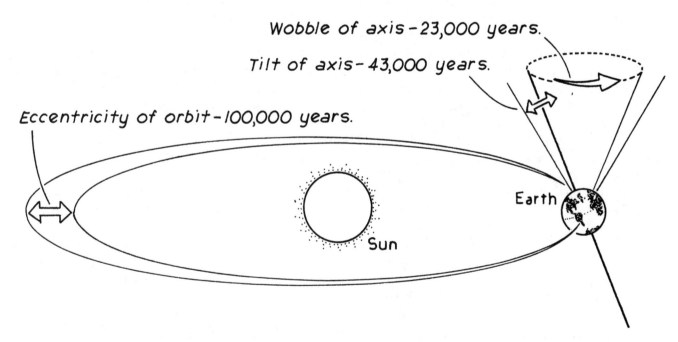

Wobble of axis – 23,000 years.

Tilt of axis – 43,000 years.

Eccentricity of orbit – 100,000 years.

Earth

Sun

Variations in Earth's Orbital Features

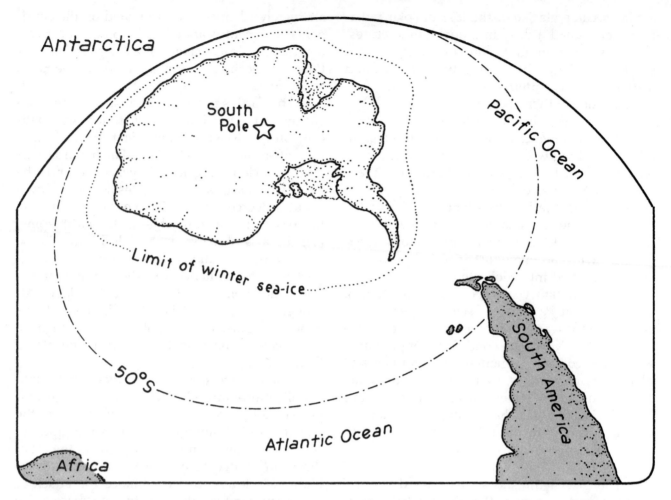

Antarctica

South Pole ☆

Pacific Ocean

Limit of winter sea-ice

50°S

South America

Atlantic Ocean

Africa

have been proposed which might amplify small changes in temperature or precipitation.

There may be strong links between the northern glaciers and the mysterious ice-covered continent at the bottom of the globe. We tend to think of Antarctica as "out of sight, out of mind." But the huge volume of water stored as ice on that huge continent must surely have a significant effect on world climate. There is ten times more ice on Antarctica than on Greenland!

Much of Antarctica has been depressed below sea level by the weight of the ice. The depression of the continent exposes the base of the ice to attack by the sea. Several investigators have argued that the Antarctic ice sheet is inherently unstable. The base layer of the ice is warmed by heat flowing from the interior of the Earth. When the ice reaches a certain critical thickness, the pressure of the overlying mass can bring the base layer to the melting point. Lubricated at the base by its own meltwater, the continental ice can begin to slide seaward. The slide is resisted by friction, which generates more heat and more melting. More melting facilitates the slide. The process is self-

accelerating. A great part of the continental ice cap can collapse toward the sea in a catastrophic surge.

An Antarctic ice surge would raise sea levels and create a large shelf of sea ice around the southern continent. The shelf might extend as far north as latitude 50 degrees south, beyond the tip of South America. Ice reflects back to space more incoming solar radiation than water. The sea ice resulting from an ice surge might decrease the heat input to the Earth by as much as 4 percent. This in turn would initiate glaciation of the northern continents and contribute to a further reduction of solar heat.

There is evidence for an Antarctic ice surge at about the time of the beginning of the last pulse of northern glaciation. Some geologists have suggested that periodic Antarctic ice surges might have been the trigger for recent advances of the ice on the northern continents.

As the sea ice resulting from an Antarctic ice surge breaks up and melts, the situation returns to normal—until the Antarctic ice cap again builds up to the critical thickness for a new surge. Then the cycle is repeated. Some scien-

The Iceman Cometh

tists believe that even now part of the Antarctic ice sheet may be approaching the point of instability. A new surge might happen at any time and plunge the Earth into another ice age.

Ice ages cause dramatic changes in habitats for life. Sea levels are lowered when water is stored on continents, drying out many shallow-water environments. Landmasses are made uninhabitable by ice. Cooling climate, changing patterns of circulation of air and water, all pose precarious new circumstances for plants and animals. Some species have thrived on the harsh conditions. Others have perished.

The ice ages posed dangers and opportunities for life on Earth. Some creatures faced extinction. For humans, the crisis evoked an explosion of creativity.

In 1940, four boys exploring a hole in the ground at Lascaux, France, discovered the most spectacular gallery of ice age art yet found anywhere on Earth. Engravings and beautifully colored paintings of horses, bulls, and other figures adorned the walls. Lascaux Cave is the Sistine Chapel of the ice age.

When I heard this story in the 1940s, I thought it the most marvelous thing that could happen to a boy. I dreamed of finding a similar cave beneath my backyard in Chattanooga, Tennessee. Later, more realistically, I harbored a dream of visiting the cavern at Lascaux.

By the time I had the independence and resources for travel, the cave had been closed to the public. Bacteria and mineral salts brought into the cavern by thousands of daily visitors had promoted the growth of an algae on the walls of the cave. The microscopic invaders threatened to destroy in a matter of months paintings that had maintained their hidden brilliance for 20,000 years. The growth of the algae was soon checked with antibiotics, but the gallery has remained closed to the general public.

On my first trip to Europe I headed for the next-best alternative to Lascaux, the cave at Altamira on the northern coast of Spain. Though not as extensive as the gallery at Lascaux, Altamira contains equally brilliant renderings of bison (see left), horses, deer and boars. The animals are painted on the ceiling of a cavern only 4 to 5 feet high. I had to lie on my back to view the cavorting animals. The experience was one of the most memorable of my life.

When the Altamira paintings were discovered in 1879, few people—including scientists—were willing to accept these magnificent works of art as the creations of "cave men" who had lived tens of thousands of years ago. Paleontologists had only recently begun to tell the story of an early human presence in Europe. The idea of an ice age was itself very new. That humans sheltering in Spanish caves from the rigors of an arctic climate could have created works of such delicate beauty seemed to defy belief. Acceptance of the true nature of the paintings came slowly.

The name given by paleontologists to the people who decorated the caves at Lascaux and Altamira is Cro-Magnon. The name derives from a rock shelter not far from Lascaux where skeletal remains of the cave people were unearthed in 1868. The discovery came less than a decade after Darwin's *Origin of Species*. People were still getting used to the fact that the human race was more than a few thousand years old.

It is now more than a century since the term "Cro-Magnon" entered the scientific vocabulary. In the intervening years paleontologists have pieced together a detailed timetable of human evolution. At the same time, earth scientists have constructed a timetable of the changing climates of the past. It is now clear that the two chronologies are linked and that the flower-

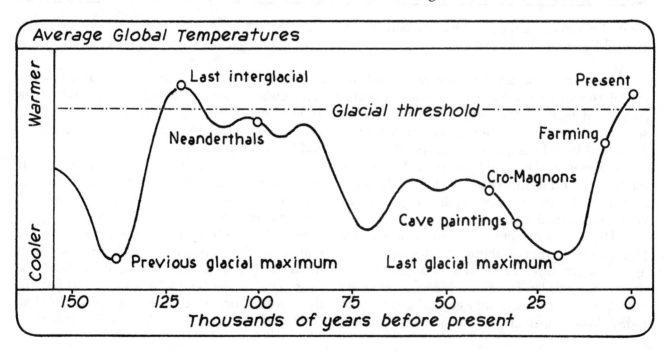

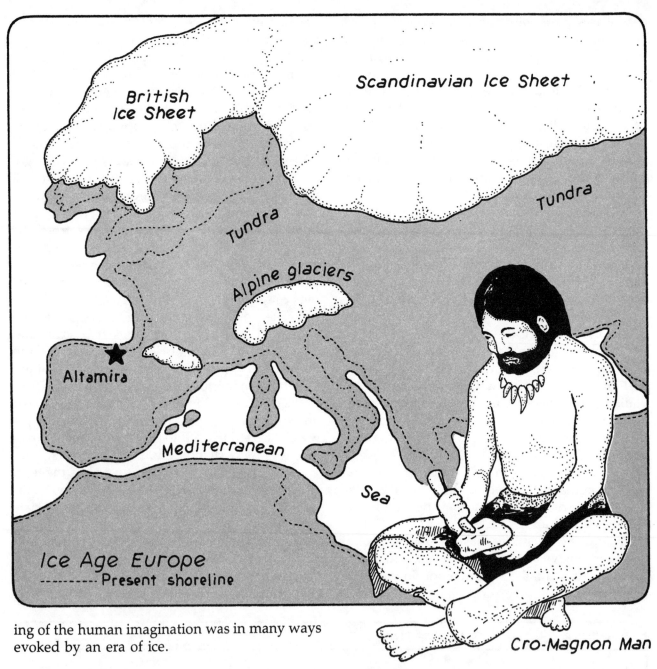

British Ice Sheet

Scandinavian Ice Sheet

Tundra

Tundra

Alpine glaciers

Altamira

Mediterranean

Sea

Ice Age Europe
- - - - - - - Present shoreline

Cro-Magnon Man

ing of the human imagination was in many ways evoked by an era of ice.

The Earth's climate has hovered about the glacial threshold for the past 3 million years. Again and again great sheets of ice have pushed south from northern centers of accumulation to scour and bury large parts of Europe, Asia, and North America. The graph on page 149 shows the last 150 thousand years of climatic history, including the most recent cycle of glacial advance and retreat.

The northern continents were mostly ice free for a brief interval about 125 thousand years ago. Glaciers retreated to high mountain valleys or remote arctic lands, sea levels rose around the world, and the climate was generally similar to today. Warm northern latitudes invited human migration from more southerly habitats. This comfortable interlude lasted about 10,000 years. Then temperatures again began to fall. Ice accumulated on Canada, Scandinavia, and Siberia and began to push south.

The initial stages of glacial advance coincided with an expansion of Neanderthal humans into new habitats worldwide. Neanderthals were precursors of the Cro-Magnons and represented the product of millions of years of primate evolution. They were sufficiently like us in mind and body to deserve the name *Homo sapiens*. They knew how to use stone and fire and they were successful hunters. Neanderthals were well established in Europe 100,000 years ago. Their success was based to a great extent on an innovative stone-tool technology. Neanderthals learned how to strike sharp-edged bladelike flakes from nodules of flint. The method was fast and efficient and led to a variety of special-purpose tools.

Skillful hunting and intelligence enabled Neanderthals to survive on the frozen tundra near the edge of the great northern glaciers. But the pressure of advancing ice posed formidable challenges. The ice was not the only threat to Neanderthal culture. About 40,000 years ago the Neanderthals were replaced by the more intelligent and fully modern Cro-Magnons. Whether Cro-Magnons were an evolutionary offshoot of the Neanderthals or a parallel race remains unsettled.

Forty thousand years ago Cro-Magnons lived in the same caves in Spain and southern France that had previously sheltered Neanderthals. The new race demonstrated a remarkable adaptability to a variety of environments. They migrated to all habitable regions of the globe (see next chapter). As the ice sheets pushed farther south, Cro-Magnons quickly adapted to life on the tundra. They developed new technologies. They mastered the production of tools from stone and bone. They learned to clothe and house themselves in ways that effectively warded off the cold. They built hotter fires. They hunted birds and fish and mammals with a skill that far surpassed the slower-witted Neanderthals.

The advancing ice may have been part of the environmental pressure that evoked new levels of human intelligence. Spoken language leaves no fossil, but ice age Cro-Magnons created a variety of enduring artifacts that speak eloquently of a subtle life of the imagination.

The Neanderthal flint blade or chopper was a remarkable achievement in its time. It was surpassed by the blademaking skills of the Cro-Magnons. Cro-Magnons learned to put an edge on stone that could rival a modern steel knife. Their implements of bone were often decorated with delicately rendered figures of the hunt. Perhaps even more interesting are the notches that are often found on their bone implements. Some archeologists believe the notches are a counting notation, like the notches on a gun. The notches have been taken to record phases of the moon, time intervals, or other countable features of the environment. Whatever their true purpose, the markings on Cro-Magnon bone implements give evidence of a rich cultural expression far in advance of mere toolmaking. The total achievement of Cro-Magnons, including cave painting, sculpture, toolmaking, and the beginnings of mathematics, timekeeping, and art, must surely represent one of the golden ages of human history.

As I lay on my back in the cave at Altamira, dazzled by the brilliant panoply of sensitively rendered animals cavorting on the roof of the cavern, it became clear to me how intimately the lives of ice age humans were bound up in the web of nature. Earth, sky, plants, animals, wind, water, ice, and fire were experienced by these people as a seamless fabric. It is a perception that is difficult to retrieve today.

The Cro-Magnon economy in Europe rested upon herds of animals that ranged the frozen tundra and swam in the cold streams and lakes at the edge of the ice. Animals supplied food, clothing, shelter, and tools. The Cro-Magnons observed keenly the habits of the animals they hunted. They translated their observations into successful hunts—and into works of art of startling beauty and fidelity to nature.

Many of the animals hunted by Cro-Magnons are familiar. Horses, deer, salmon, seals, chamois, and reindeer were among their quarry. But other creatures of the ice age have vanished, such as the giant deer (sometimes called the "Irish elk"), the wooly mammoth, and the cave bear. The extinction of these magnificent animals was part of a worldwide wave of extinctions that accompanied the waning of the last ice age. The extinctions were most dramatic in North America (see next chapter), but they can be traced on every continent.

The large mammals that inhabit the Earth today are only a pale shadow of the remarkable variety of beasts that ranged the continents only one million years ago. Not even the zoological richness of an African game park approaches the preglacial fauna. It was a Brobdingnagian fauna dominated by giants, mammoths twelve feet high at the shoulders with tusks equally as long, bison with hornspans that exceeded a man's reach, ground sloths as heavy as elephants, pigs the size of rhinos, sheep that stood as tall as horses, rodents as big as calves.

Most of the giant mammals that roamed the Earth a million years ago survived the trauma of

Extinct Animals
of the Ice Age

Giant Deer

Wooly Mammoth

Cave Bear

repeated onslaughts of glaciation, only to vanish with surprising suddenness about 10,000 years ago. The cause of this wave of death has been hotly debated and there are no certain answers. Theories for the ice age extinctions tend to fall into two categories. The first group of theories assumes the deadly influence of human predators.

The extinctions at the end of the last ice age are contemporary with the development of new and efficient hunting technologies. Humans hunted in highly organized social groups, multiplying their effectiveness with speech and reason. They used an impressive arsenal of lethal weapons, including fire. They learned to stampede herds of animals over cliffs or into enclosed slaughtering grounds. Human hunters may well have killed in excess of their needs for ritual reasons or for the pure joy of sport. And unlike predators such as wolves, human hunters sought out the strongest members of a herd rather than the weakest, again perhaps for reasons of ritual or sport.

In favor of the theory of human overkill is the fact that sea animals and small land animals seem to have escaped the wave of death. Sea animals had a ready escape from human predators. Small mammals might have escaped extermination by virtue of greater numbers and a more rapid breeding rate.

Not all paleontologists are willing to admit that small nomadic bands of hunters could have wreaked so complete a devastation. Many of the animals which were most extensively hunted by Cro-Magnons are among the animals which survived the extinctions. Many animals which were undesirable as human prey became extinct. These observations have led some paleontologists to suggest that climate was the culprit. Certainly the time of the extinctions was a time of rapidly warming temperatures that followed the last glacial maximum. Large land animals would have been least likely to adapt successfully to the new conditions.

But we now recognize that the last glacial cycle was only one of many that have affected the Earth over the past 3 million years. If climate was the cause of the wave of extinctions, then why did animals survive previous episodes of cooling and warming only to fall victim to the most recent one?

Perhaps the true cause of the extinctions 10,000 years ago has not yet been discovered. It is not impossible to rule out an external catastrophe, such as a nearby supernova or meteorite impact, as the instrument of extinction. All that can be said with certainty is that the animals I have sketched at left and many like them, disappeared from the Earth at a time when a cool, moist climate was changing to a warm, dry one. It was also a time when humans in expanding numbers, spurred by the challenge of survival in an ice age world, turned keen intellectual resources to the technology of killing.

The Most
Terrible Tiger

Sea-level changes during the most recent ice age created a land bridge between Asia and Alaska and opened the Americas to human migration. The effect on the native animal population was catastrophic.

There are no examples of ice age art in the Americas comparable to the magnificent cave paintings at Lascaux and Altamira. I have reproduced here the only object of American art from that period, a rendering of two animals on a piece of bone. The engraving shows an elephantlike mastodon and what might be a large cat. The object was found in Mexico and appears to be 22,000 years old.

The engraved bone from Mexico is one of the few shreds of evidence to suggest a human presence in the Americas before 12,000 years ago. All of the evidence for an entry into the western continents before that time has been disputed. If humans did reach the Americas before 12,000 years ago, they were few in number, left few artifacts, and had only a minor effect on the environment. It is not certain when the first humans arrived in the western hemisphere, but it seems clear that the first arrivals came from Asia and migrated across the Bering Strait.

The Bering Strait between Siberia and Alaska is only 66 miles wide. It is possible to see the shining snowcapped peaks of one continent from the shores of the other. Humans might have passed from one continent to the other by boat or by walking across pack ice on the occasionally frozen strait. But the migration from Asia to the Americas was made easy—indeed likely—during the ice ages.

As ice built up on the northern continents sea levels dropped worldwide. Continental margins which are today submerged by the sea were then exposed. The Bering Strait between Siberia and Alaska is one place where the continental shelf is broad and only lightly flooded. As frozen water piled up on the continents and sea levels fell, Asia and North America were connected by dry land (see map). There is very little

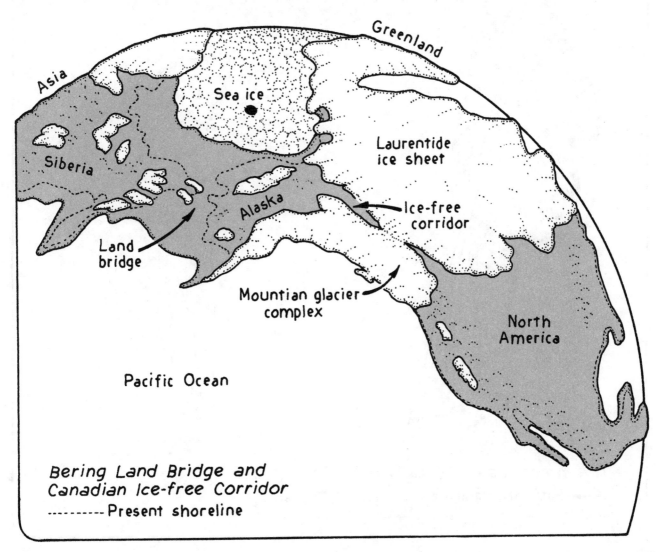

Bering Land Bridge and Canadian Ice-free Corridor
---------- Present shoreline

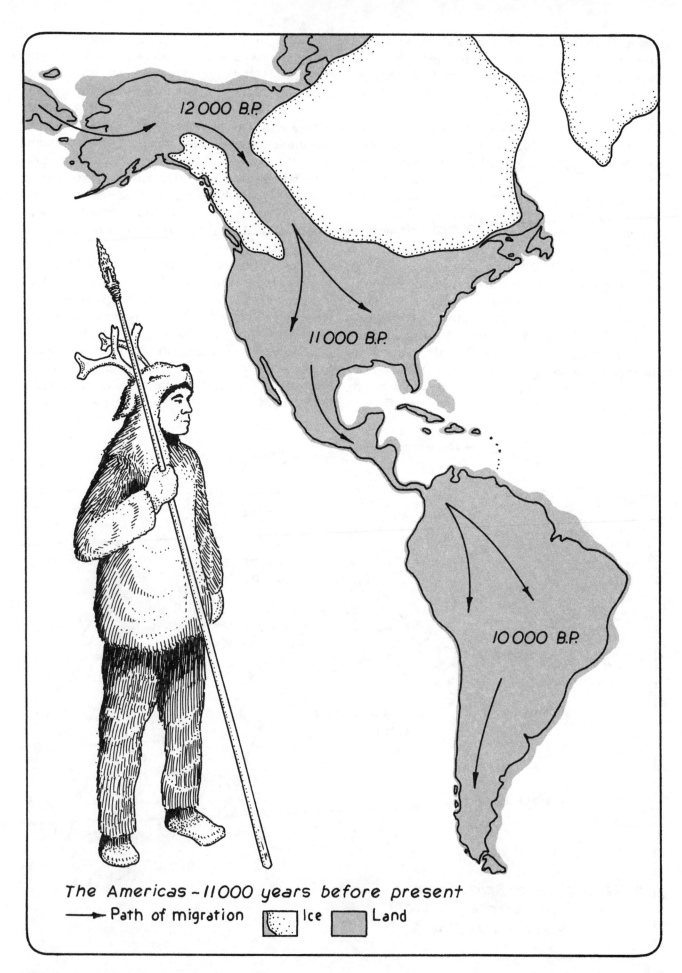

The Americas – 11000 years before present

→ Path of migration ⬚ Ice ▨ Land

12 000 B.P.

11 000 B.P.

10 000 B.P.

moisture in the air in this part of the world and so very little snow. Alaska and Siberia and the land bridge connecting them remained mostly ice free even at the height of the ice ages.

Meanwhile, ice accumulated on North America in two great sheets. One sheet, the Laurentide ice sheet, covered the eastern part of Canada and the northeastern and north-central United States. The other sheet formed as a coalescence of alpine glaciers in the western mountains. At the peak of the ice ages, eastern and western ice sheets merged to form a plateau of ice reaching from coast to coast. Migrants who walked across the Bering land bridge into Alaska were blocked by this frozen barrier from access to the rest of the continent.

When the ice sheets melted, the seas rose and flooded the Bering land bridge. Only at certain intermediate times, as the ice sheets were growing or retreating, was there sufficient continental ice to afford a dry shod passage to and from Asia, and yet not enough ice to close the corridor through Canada between the eastern and western ice sheets. The timetables of dry bridge and open corridor determined the schedule of possible migrations between the continents.

The land bridge and ice-free corridor opened and closed many times during the past 3 million years, in step with pulses of glaciation on the continents. Across the bridge and along the corridor moved a great traffic of animals, travelling both ways. Asian species, such as the mammoth, the bison, and the sabre-toothed tiger, made their way to the Americas. A smaller traffic, including the fox and the ground squirrel, went the other way.

Humans were part of the traffic entering from Asia. The most successful wave of human migration across the Bering land bridge, and the only one for which the evidence is overwhelming, occurred at the peak of the most recent glaciation. Following herds of grazing animals, nomadic tribes of hunters entered Alaska from Asia sometime before 20,000 years ago. There they confronted an unbroken wall of ice a mile or more high blocking further movement to the south. Only when the ice began to melt back about 12,000 years ago did the bottled-up Alaskans have access to the south. When the Canadian ice-free corridor opened up, animals and their pursuers poured through. Within one or two thousand years they had reached the southern tip of South America. The dates on the map give a very rough schedule of the southerly advance.

It is uncertain whether the wave of immigrants that swept through the Americas between 12,000 and 10,000 years ago encountered a prior human presence on the continents. We shall perhaps never know whether the encounter, if it occurred, between the earlier arrivals and the new invaders was friendly or hostile. What is certain is that the invaders were armed with an impressive new technology—spears tipped with gracefully fluted flint points, and spear-throwing devices that enabled a hunter to drive his spear deep into the body of his prey. The tribes equipped with these deadly weapons are called Clovis people, after a place in New Mexico where their kill sites and camps were unearthed in the 1930s. The Clovis dig turned up the bones of mammoths, horses, and bison, in conjunction with the deadly fluted points. Subsequent excavations have uncovered the characteristic Clovis points throughout the Americas. It took only a thousand years for the Clovis people, armed and ready for adventure, to subdue two continents.

It surely required enormous courage for a nomadic people to traverse the ice-walled corridor through Canada. At times the passage may have been less than a hundred miles wide. Only the pursuit of migrating mammoths, mas-

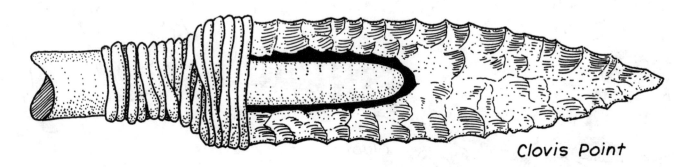

Clovis Point

todons, or other big game could have motivated bands of hunters and their families to enter the icy channel. Once south of the ice, the invaders found themselves in a fertile, undreamed-of Eden, populated by an extraordinary range of abundant animal life.

The ice age fauna of the Americas, as in Europe, was dominated by giants. There were great wooly mammoths and lumbering mastodons with long recurved tusks. These elephantlike beasts ranged from the Arctic to Patagonia. There were sabre-toothed "tigers," cats large enough to subdue a mammoth. There were sloths as big as rhinos and beavers the size of bears. Along with the giants, great herds of horses, bison, and camels roamed the grasslands. The hunters who encountered these vast new continents so rich in ready game must have deemed them paradise.

The encounter of humans with paradise was devastating—for the natives of paradise. Within a brief thousand-year interval, the large ice age fauna of North and South America was decimated. The sudden dying-out of the great beasts has long been obvious to paleontologists. Even Darwin, who explored for fossils along the coasts of South America in the 1830s, commented on the sudden transition in the fossil record from "monsters" to "pigmies."

Not all scientists agree that the entry of humans into the western hemisphere was the cause of the extinctions. As in Europe, a case can be made for changing climate as the culprit. But the evidence linking human overkill to the disappearance of land animals is more compelling in the Americas than in the Old World. The wave of human migration can be traced through finely crafted flint points down across the western continents from the Canadian arctic to Tierra del Fuego. Often the deadly spear tips are found in conjunction with the bones of dead beasts. The native animals of paradise recoiled before the human invasion. For these animals, the western hemisphere became a trap, a cul-de-sac, a slaughtering ground with no escape. As one scientist put it, "Man was the most terrible tiger."

The advancing tide of human hunters

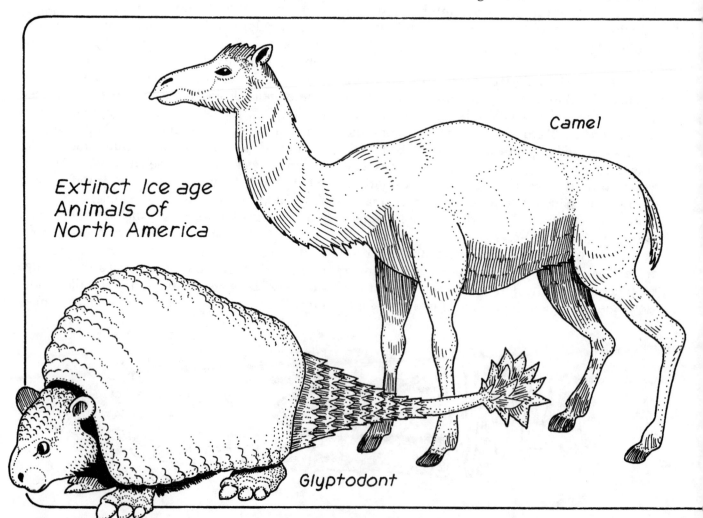

Extinct Ice age Animals of North America

Camel

Glyptodont

158

swept south at a pace of perhaps ten miles per year, doubling in numbers every generation. Within a few hundred years after emerging from the ice corridor, the wave had reached the Gulf of Mexico. Behind lay the bleaching bones of tens of millions of large animals. There was a pause, perhaps, at the bottleneck of Panama. Then the wave burst into South America and caused further devastation.

On these pages I have sketched a few of the animals that vanished from the Americas at the end of the last ice age. They offer a glimpse back across a great divide in evolutionary history, the transition from the Age of Mammals to the Age of Humans.

The armadillolike glyptodont was one of the South American animals that successfully moved north across the Panama land bridge. The beast was as heavily armored as a tank. But it was no match for the human hunter who could flip it over and plunge a flint-tipped spear deep into its unprotected belly.

Camels and horses were natives of the Americas that migrated to the Old World across the Bering land bridge. They fared better in their new environment than in their native Americas. Not long after the arrival of humans, American horses and camels had been hunted to extinction. The wild horses that roam the west today are descendants of stock that arrived in Spanish ships at the time of the Conquistadores.

The scourge of the native American mammals was the sabre-toothed "tiger." This sleekly efficient cat clutched its victim with massive forelimbs and paws and stabbed with daggerlike upper teeth. The fierce sabre-tooth shared the Americas with some other deadly killers. The dire wolf was larger than today's timber wolf and was armed with bone-crunching jaws. The American panther was bigger than a lion. All of

Giant Ground Sloth

Sabre-toothed "Tiger"

these fast, powerful carnivores fell before the most deadly predator of all, the two-legged hunter equipped with intelligence, speech, fire, and flint-tipped spears.

The giant ground sloth was another North American which came from the south. Its technical name, *Megatherium*, means "great beast." This ponderous, slow-witted giant could nibble leaves off branches 20 feet above the ground. It had long curved claws capable of pulling down even higher branches or digging up roots and tubers from the ground. The ground sloths were exceeded in size only by the mastodons and mammoths. The sloth shared the fate of the larger beasts. Three species of mastodons and four species of mammoths vanished with the sloth. Rhinos, bears, tapirs, and peccaries went with them.

And so an era came to an end. When the reptiles lost their hold on the planet 63 million years ago the mammals moved into the ascendency. When the ice ages began about 3 million years ago mammals occupied almost every available ecological niche on the planet. And like the reptiles, the mammals had given rise to a race of giants. Mammoths and ground sloths were the mammalian "dinosaurs." Then, within a thousand-year period at the end of the last ice age, the giants fell, and the "Golden Age of the Mammals" came to a close.

The cause of the wave of extinctions is debated, but there is little doubt that humans were a factor in the demise of the giants. With skills honed on ice, one mammal moved into an ascendency above all others. The success of the new champion was not based on size, or armor, or sabrelike tooth and claw, but on intelligence. The "Golden Age of Mammals" gave way to the "Age of Humans."

The rigors of the ice age evoked in the human mind a wonderful flowering of imagination. At the very height of the glaciations we find in the caves of southern Europe magnificent works of art that have seldom been surpassed. And in the decorated artifacts of the cave inhabi-

tants we find evidence for the beginnings of religion, science, and technology.

With intelligence and imagination came access to deadly force. Humans learned to turn fire and flint to the business of survival. No longer would nature require millions of years to perfect tooth, or claw, or scale, or feather. No longer would size or swiftness, or armor ensure survival. From the end of the ice age forward the scale and pace of planetary change would be shaped more by human craft than by the patient random shuffle of the genes.

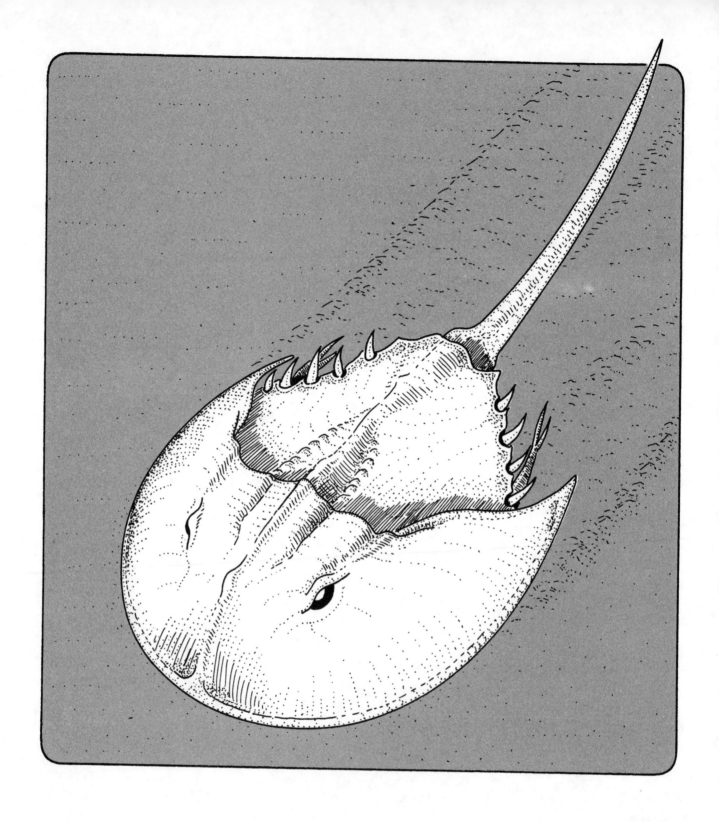

Staying
the Course

Some animals and plants have endured almost unchanged through hundreds of millions of years of geologic and cosmic violence. Their story illustrates new ideas about evolution.

Evolution is the story of change. It is a story of adaptation to a changing environment and of a tendency toward complexity and intelligence. Survival is the name of the game. Life spurns no trick, withholds no sleight of hand, to maintain its hold on the planet.

This book has been mostly about the response of life to geological and cosmic violence. It has been a story of novelty and extinction in response to stress. New strategies for survival have appeared in response to dangerous alterations of the environment. Whole families—the trilobites and dinosaurs, for example—have failed to make the grade and have passed away. In this chapter we will look briefly at creatures that have resisted change in the face of stress and have nevertheless endured. And we shall consider some new ideas about evolution.

Let's begin with the fellow at left, hauling himself across the sand near Chesapeake Bay. This tin can with a spike is the horseshoe crab.

The horseshoe crab—or its empty shell—is a familiar sight on Atlantic beaches. Almost identical fossils have been found in rocks as old as the Cambrian. The horseshoe crab is one of the earliest actors to appear on the stage of multicellular life. For 500 million years the tanklike invertebrate has survived every threat nature has hurled against it. We marvel at its staying power, and wonder what stickiness of genes enabled it to resist the transforming push of evolution.

Genes may be stickier than we used to think. The tendency of life to resist change may sometimes be strong or stronger than the tendency toward adaptation. The clams and the brachiopods are a case in point.

The clams and the brachiopods have long been a textbook case for the dynamics of evolution. The two groups share the same habitats. They have similar lifestyles. Certain members of the two groups even look alike. And it seemed from the fossil record that a continuing expansion of the clam's domain had been accompanied by dwindling success for the brachiopods.

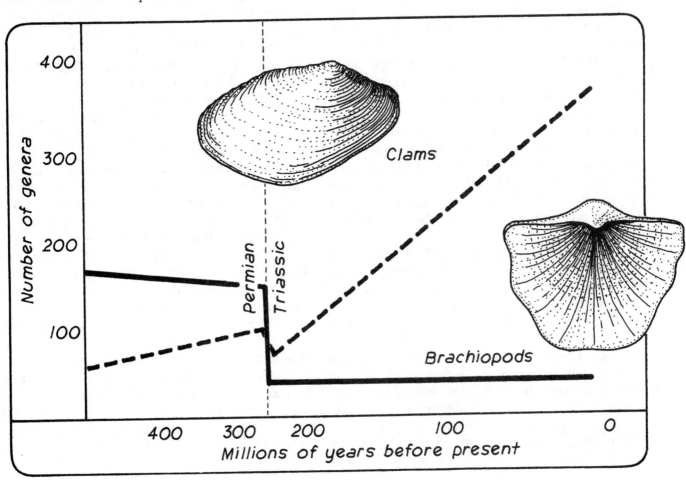

Classical neo-Darwinism provides a ready explanation for the closely related histories of the two groups. The clams and the brachiopods compete for the same ecological niche. The success of one group must come at the expense of the other—nature red in tooth and claw. The keen competitive race for survival, acting over the eons, had selectively favored the better adapted clams over the brachiopods. The clams showed greater resourcefulness, neo-Darwinists would say, and have been rewarded with success.

Or have they? A study by Stephen Gould and C. Bradford Calloway has cast doubt on this conventional wisdom. Their statistical survey of the number of genera of clams and brachiopods across geologic time is summarized by the graph at the bottom of the previous page, which has been adapted from their study. The lines on the graph show the number of genera across geologic time. I have smoothed the lines to show general trends only. If I had not smoothed out the data it would be clear that the minor ups and downs of one group are followed by the ups and downs of the other. There is no hint in these fluctuations that the success of one group came at the expense of the other. The groups seem to have waxed and waned together in response to environment.

Over the long sweep of geologic time, however, the story is different. The clams have shown a steadily increasing number of genera and the brachiopods have merely held their own. The race for survival was decisively won by the clams. But the long-term outcome of the race was not so much due to a competitive edge of clams over brachiopods as to one decisive event, the Great Dying, the crisis of yet unspecified character that closed the Permian era and began the Triassic.

Two hundred and thirty million years ago, half of the existing families of marine organisms vanished from the Earth. Land plants and animals were also disrupted, although not so decisively. Paleontologists are not certain of the cause of the Great Dying. It might have been the ecological consequence of the stitching together of the continents into the seamless garment of Pangaea. It might have been instigated by an asteroid impact or a nearby supernova or massive volcanic activity or any of the other calamities we have considered in this book. For

the moment, the cause is not so important as the effect.

The clams passed though the crisis relatively unscathed. Perhaps a third of all clam genera became extinct. The brachiopods suffered a heavier blow. Their numbers were chopped by three-quarters. The subsequent histories of these two clans can be read in their response to the single devastating blow of the Permian-Triassic crisis. The trauma "reset the clocks" for the two groups. The "clock" of the brachiopods was set back further than the "clock" of the clams. When the clocks got going again the clams had gained an edge and there was no looking back.

Classical neo-Darwinism stresses gradual evolution in response to the pressures of competition and environment. The study of Gould and Calloway, and many other recent studies, suggest that there have been long periods of stability within the record of life, punctuated by episodes of rapid change. According to the new ideas about evolution, the story of life unfolds like a slide show rather than a movie. Single frames linger on the screen for a long time and change with sudden and clearly perceptible jumps. The older theory saw evolution rolling forward like a movie, a progression of many brief frames that merge one into the next with but slight difference. Classical neo-Darwinism focused on the pressures for continuous and gradual evolutionary change. The new agenda for evolutionists is to explain the constraints that kept things constant through epochs of time.

Evolutionists have long been troubled by the problem of "missing links." The fossil record is replete with examples of organisms that differ greatly from their ancestors, but without a record of intermediate variations. If evolution proceeds by slight increments, as Darwinism requires, then somewhere in the record of the rocks there should be fossils delineating every step of the transformation.

Darwinists have taken refuge in the incompleteness of the fossil record, and with considerable justification. The book of nature, recorded in the rocks, is marred by many missing pages. Just as the deposition of sediments lays down the pages of the book, erosion at other times rips them away. There are plenty of gaps in the stratified rocks in which missing links could

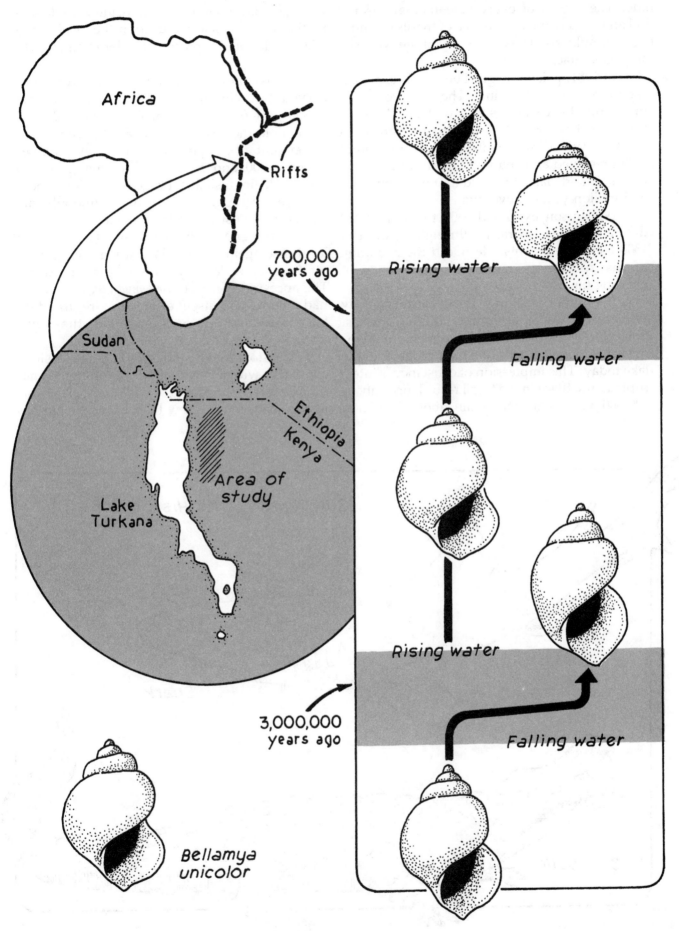

Africa

Rifts

700,000
years ago

Sudan

Ethiopia
Kenya

Area of
study

Lake
Turkana

Bellamya
unicolor

Rising water

Falling water

Rising water

Falling water

3,000,000
years ago

hide. The dream of every paleontologist is to find an uninterrupted sequence of fossils extending over millions of years, a story of life with all the pages intact.

In 1981, paleontologist Peter G. Williamson reported just such a find. The discovery was made on the eastern shore of Lake Turkana (formerly Lake Rudolf) in northern Kenya. In neatly stratified sediments, Williamson found well preserved and abundant fossils of freshwater snails and clams that told an unbroken story of 3 million years of evolution.

Williamson examined 3300 fossils from 13 different species, extracted through a depth of 1300 feet of sediments. He traced the evolution of his specimens across broad reaches of geologic time.

The most striking thing observed by Williamson was the stability of the species. Some fossils extracted from sediments millions of years old look exactly like their progeny in the lake today. This impression of constancy is interrupted only twice in the fossil record, once about 700,000 years ago and again about 3 million years ago. Both disruptions coincide with deposits that record a sudden drop in the level of the lake, undoubtedly precipitated by changing climate.

I have illustrated the general trend observed by Williamson with just one of his species, the freshwater snail *Bellamya unicolor*. At each fall in the level of the lake *Bellamya unicolor* underwent a brief period of change, lasting between 5,000 and 50,000 years, giving rise to clearly distinguishable progeny species. These episodes of rapid evolution were undoubtedly caused by the stress of falling water.

At the end of the dry intervals, the water level rose again and the lake was no longer cut off from other lake systems in eastern Africa. The ancestral species of *Bellamya unicolor*, which had maintained itself elsewhere, returned to Lake Turkana. At the same time, the novel species disappear from the fossil record.

The story told by the Lake Turkana fossils is consistent with a new theory of evolution sometimes called *punctuated equilibrium*. The difference between the new theory and the old has

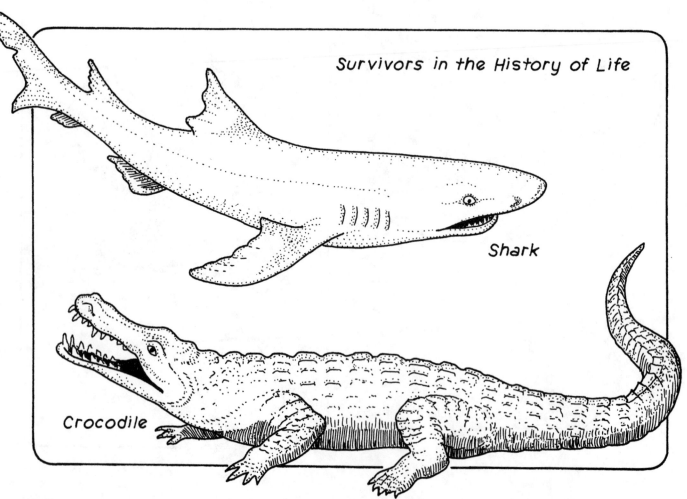

Survivors in the History of Life

Shark

Crocodile

nothing to do with the *fact* of evolution or with the *genetic mechanism*, but with the *tempo* of change. The new theory, like the Lake Turkana fossils, emphasizes long periods of constancy punctuated by brief stress-induced episodes of instability giving rise to new species. In the case of Lake Turkana, the stress that instigated the rapid evolution of new species was a catastrophic lowering of the level of the lake.

The debate about these matters—about the tempo of evolution—continues to rage within the biological community. Theories of punctuated equilibrium seem to be finding increasing favor. It may be that there are evolutionary constraints at work within living things that make certain kinds of developmental changes difficult or impossible. According to this view, only dramatic external disruptions of these constraints allow the development of new species.

The history of life and the history of the planet are closely linked. The tempo of evolution has been orchestrated by geological and cosmic violence. The drift of continents, upheavals of climate, rising and falling sea levels, crescendos of volcanism, the deaths of stars, and the crash of meteorites have all served to punctuate the story of life. Meanwhile, life has done its best to stabilize its habitat. We should not underestimate the ability of even the simplest organisms to modify the global environment to their own liking.

The new theories of evolution emphasize the constraints in life that tend to keep things the same. If the Earth had been a steadier platform for the pageant of life, then perhaps the pageant would have been a dreary bore. Catastrophes in the physical environment have induced new species and caused the extinction of old ones. The texture of the drama of life, the complexity of the plot, the variety of the characters on the stage, may well be the artifacts of violence.

Most of this book has been devoted to novelty and extinction, to the grand episodes of terrestrial catastrophe and evolutionary change. On these last few pages I parade for your appreciation some of those creatures that have managed to maintain themselves across re-

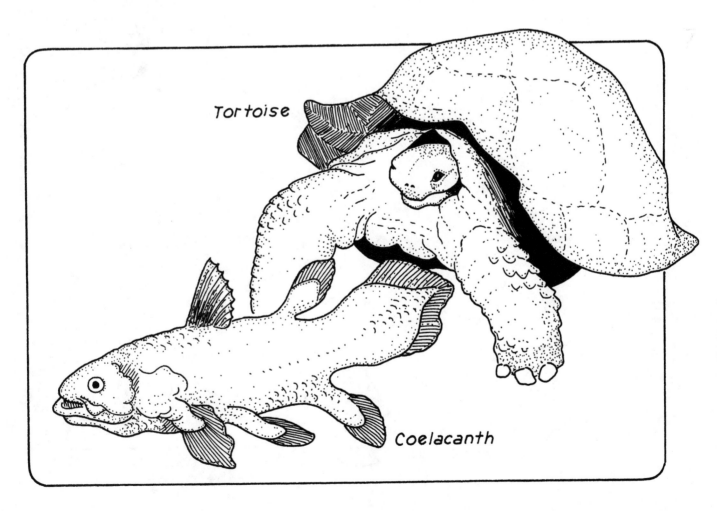

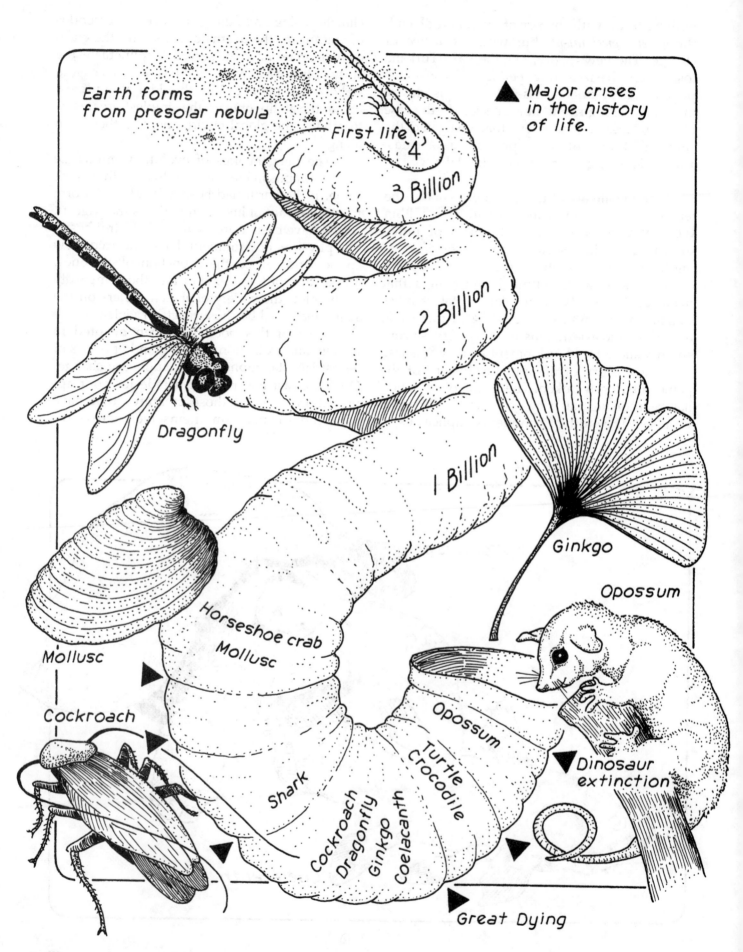

Earth forms from presolar nebula

▲ Major crises in the history of life.

First life 4

3 Billion

2 Billion

Dragonfly

1 Billion

Ginkgo

Opossum

Mollusc

Horseshoe crab
Mollusc

Cockroach

Shark

Cockroach
Dragonfly
Ginkgo
Coelacanth

Opossum

Turtle
Crocodile

▲ Dinosaur
extinction

▲ Great Dying

peated crises in the history of life. Like the horse-shoe crab, each of these animals or plants is a kind of living fossil, a throwback to an earlier time, an antique in nature's attic. They represent the staying power that seems to be built into life. They evoke the steady gene that hankers to stay the same. These are the creatures that have tenaciously hung on when the world went topsy-turvy.

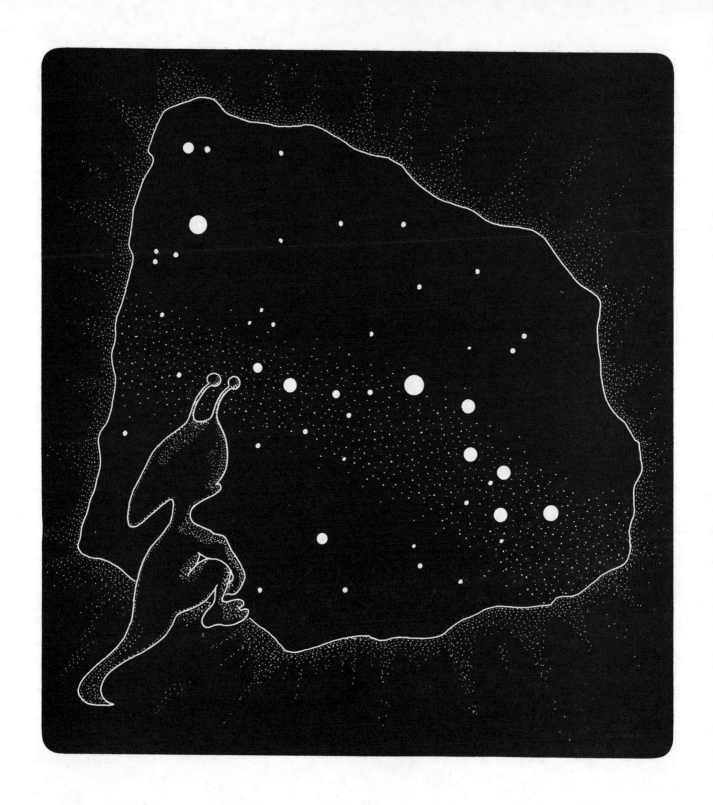

The Ultimate
Catastrophe

One animal on Earth has acquired the power to change the environment on a scale far beyond geologic and cosmic catastrophes of the past. This great power carries a concomitant responsibility.

I would like to open this last essay with a view from a far-off place. Accordingly, I asked my son to design for me an extraterrestrial and I have placed the little fellow at the mouth of a cave on a planet of a star in the system Alpha Centauri. Let us imagine that evolution on this Centaurian planet has brought my little E.T. to the threshold of intelligence and wonder—to the point, say, our human ancestors were at a hundred thousand years ago. My extraterrestrial sits at the mouth of its shelter and looks out at the starry night. Both Centaurian suns have set, the yellow sun and the orange sun. The night is dark and brilliantly spangled. The Milky Way sweeps across the sky like a sash of feeble light. The W-shaped group of five stars known on Earth as Cassiopeia burns brightly at the lower right. At the upper right is the dazzling yellow star Earthlings call Capella. Between Cassiopeia and Capella, almost centered in the view from the mouth of the cave, is another yellow star, almost a twin of Capella. It is a star that has never been seen in the night sky of Earth. It is the Earth's daytime star. It is the Earth's sun.

Earthlings don't often think of their sun as a star. That godlike burning globe seems to have little in common with the cold points of light in the night sky. But at the distance of Alpha Centauri—4.3 light-years—the Sun is just one more nighttime spark of light. It is not even the brightest star in the Centaurian night. It floats serenely in the stream of the Milky Way, just one of the hundreds of billions of stars in that grand spiral galaxy.

In the view from Alpha Centauri the planet Earth has shrunk to invisibility, a speck of dark stone absorbed in the Sun's greater light. Perhaps my extraterrestrial wonders if out among the stars there are other intelligences like its own. As it contemplates the immensity of the cosmos, perhaps it wonders if it alone imbues the Universe with consciousness and life.

No moment on any planet anywhere in the Universe can be more special than the moment of first life. Perhaps the appearance of life happens only as a freak accident, as a once-in-a-universe chance arrangement of molecules, the roll of a die with a thousand billion faces. If that is so, then there is no Centaurian at the mouth of a cave near that other star and we are alone in the immensity of the Galaxy. Then the stars that decorate our night sky are motes of lifeless dust.

Or perhaps life is inevitable. Perhaps life is built into the very stuff of nature, singing at the heart of every atom, waiting only for the right time and the right place to spring into existence. Then, somewhere near another of night's stars another planet is alive. Then, somewhere, another creature marvels at the night sky and wonders at the mystery of life.

I will confess that I have always been inclined to the latter view. Having written this book I am even more inclined to believe that life is installed at the core of all that exists. Life appeared on Earth three and a half billion years ago. Since that time it has been buffeted by every sort of geological and cosmological violence. At times life has been its own worst enemy. And still it has flourished. It has lodged itself in every niche on the planet's surface, from upper atmosphere to ocean deep, from searing equator to arctic ice. Life has become so pervasive, so intricately entangled with rock, wind, and water that it is perhaps best to think of life and the planet Earth as one, not as inhabitant and habitat, but as a single living organism in orbit about a yellow star.

If life was inevitable, then so, no doubt, was consciousness. And so now on Earth there is an animal capable of reconstructing imaginatively the catastrophes of the past and anticipating catastrophes of the future. Inevitably we wonder if the violence of the past might recur to afflict our present. Earthquakes and volcanoes are always with us, but what of those greater catastrophes that have caused extinctions of entire species and redirected the course of evolution?

What, for instance, if the present period of relatively warm climate is only the brief interruption of a long regime of ice? On the basis of past climatic cycles there is every reason to believe that this might be so. What if glaciers again build up on Canada and Scandinavia and push south, scooping up cities and grinding them to dust. Could the ice once again reach Manhattan Island, slide down across Central Park, topple over bridges and skyscrapers?

Or what if the unrestrained burning of fossil fuels, deforestation, and changing land use upset the carbon cycle, increasing atmospheric carbon dioxide and instigating a "greenhouse effect." Could the Greenland and Antarctic ice sheets melt and flood the continental margins? A melting of all the ice presently piled up on land would raise the level of the ocean to Lady Liberty's chin. Most of the great cities of the world would be deluged. Florida would simply disappear.

These are awesome prospects, and they might well happen, but they would certainly not give rise to the theatrical scenes I have sketched on these last pages. Glaciers require thousands of years to build up on continents or to melt away and augment the seas. By the time an advancing ice sheet approached New York, the city would be a crumbling ghost town, abandoned by its inhabitants for more felicitous climates.

New cities would have sprung up further south on continental shelves now flooded by the sea. Or, in the scenario of melting ice caps, by the time the water reached the Statue of Liberty's chin, New Yorkers would have retreated to higher ground and the famous monument would have fallen into the sea.

More likely, such calamities will not happen at all. Something new has appeared on planet Earth—human technology of an awesome power. Only a modest enhancement of present technology would give humans control of the Earth's climate. Huge mirrors in space, for example, could add or subtract sunlight from the global equation. Carbon dioxide levels in the atmosphere could be artificially adjusted. Indeed, it is hard to imagine any catastrophe which na-

ture might prepare for our descendants that they could not anticipate and effectively counter.

But there is a darker side to the same coin. As so often in the past, life might be its own worst enemy. Technology might be itself a more terrible threat than anything nature can conjure against us. Present stockpiles of nuclear weapons pose a more ominous threat to life on Earth than any of the natural calamities I have treated in this book. Asteroids and volcanic eruptions, drifting continents and grinding ice, have all served to punctuate evolution. A nuclear holocaust could also punctuate the story of life on Earth. Period.

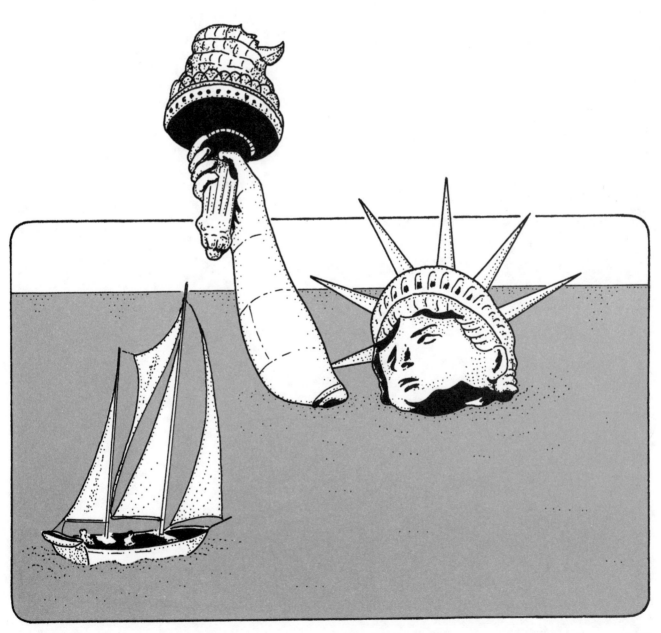

Notes and References

Page 9: For a description of the Anasazi observatory on Fajada Butte see Thomas Canby's "The Anasazi Riddles in the Ruins," *National Geographic*, 162:5, November 1982, pages 580–581.

Page 11: My drawing is based on a photograph of the Canis Major R1 region studied by W. Herbst, Van Vleck Observatory, Wesleyan University. The photograph appeared with the article "Supernovas and Star Formation," William Herbst and George E. Assousa, *Scientific American*, 241:2, August 1979.

Page 14: I have exaggerated the size of the moon and stars in my drawing of the sky over Chaco Canyon.

Page 25: The drawings on this page and on pages 30, 46, and 87 are based on photographs from the National Aeronautics and Space Administration.

Page 30: The drawing on this page is a composite. The Earth would not be so near to the lunar horizon from any of the actual Apollo landing sites.

Page 33: The drawing of a bacterium in the process of division is based on a microphotograph by S. C. Holt, Department of Microbiology, University of Massachusetts. The schematic representations of the molecular chemistry of life on this and the following pages are adapted from illustrations for R. E. Dickerson, "Chemical Evolution and the Origin of Life." Copyright © 1978 by *Scientific American*, Inc. All rights reserved.

Page 38: My drawing of *E. coli* is based on a microphotograph by S. C. Holt, Department of Microbiology, University of Massachusetts.

Page 38: The process of getting energy from sugar is more complicated than the simple net equation illustrated in my diagram. First, the sugar molecule is split into two pyruvic acid molecules by a process called glycolysis. Some microorganisms and many plants convert pyruvic acid into alcohol with the release of carbon dioxide. In animals, another fermentation pathway yields lactic acid rather than alcohol.

Page 52: Chinese records of naked-eye sunspots apparently do not confirm the activity minimum notable in European records for the period 1645–1715. See *Sky and Telescope*, September 1982, page 234.

Page 54: My drawing of the neutrino telescope is based on a photograph made available by the Brookhaven National Observatory, Upton, New York. The particular neutrinos counted in this experiment are not the ones indicated in the diagram on page 5 but are rather neutrinos created later in the fusion chain.

Page 58: The drawing of a cyanobacteria is based on a microphotograph by L. V. Leak, Department of Anatomy, Howard University College of Medicine.

Page 59: The sketch of a fossil stromatolite is based on an original photograph in "An Early Habitat for Life," Groves *et al.*, *Scientific American*, 245:4, October 1981, page 66.

Page 64: A careful naturalist will note that the two insects illustrated here would be unlikely to be on the wing in the same season.

Page 71: The drawing of *Chlamydomonas* is based on an electron micrograph made by Ursula W. Goodenough, Department of Biology, Washington University, St. Louis, Missouri.

Page 72: The whiplike appendages on eukaryotes are traditionally called flagella. Current terminology uses the term undulipodia to refer to the eukaryotic flagella and to the hairlike appendages on eukaryotic cells called cilia.

Page 73: The diagrams on this page and on page 84 are based on the work of L. Margulis of the Department of Biology, Boston University. Margulis's *Early Life* (Science Books International, 1982) is an excellent account of the early evolution of life. The definitive work on cell evolution by symbiosis is L. Margulis, *Symbiosis in Cell Evolution*, W. H. Freeman and Co., 1981.

Page 78: My drawing is based on a Department of Environment photograph of the prehistoric village of Skara Brae, with permission of the Controller of Her Majesty's Stationery Office.

Page 91: The drawing of fossil *archaeopteris* is adapted from Henry N. Andrews, Jr.; *Ancient Plants and the World They Lived In*. Copyright © 1947 by Comstock Publishing Co., Inc. Used by permission of the publisher, Cornell University Press.

Page 98: The diagram is based on J. D. Hays's "Faunal Extinctions and Magnetic Field Reversals," *Geological Society of America Bulletin*, 82, 1971, page 2433.

Page 104–105: The diagram is based on "Mass Extinctions in the Marine Fossil Record," by D. M. Raup and J. J. Sepkoski, *Science*, 215, 1982, page 1501.

Page 106: My drawing of Meteor Crater is based on a photograph in J. S. Skelton, *Geology Illustrated*, W. H. Freeman and Co., San Francisco, 1966.

Page 109: The diagram is based on C. J. Orth, J. S. Gilmore, J. D. Knight, C. L. Pillmore, R. H. Tschudy, and J. E. Fassett, *Science*, 214, 1981, page 1341. I have smoothed out the data and connected data points with a continuous line. Such procedures, while aiding clarity, can hide a multitude of sins.

Page 119: The drawing of a rabbuck is based on illustrations in D. Dixon, *After Man*, Harrow House Editions Ltd., London, 1981. The book is a delightful romp in the future.

Page 121: For a fuller account of the Nebraska dig see M. R. Voorhies's "Ancient Ashfall Creates a Pompeii of Prehistoric Animals," *National Geographic*, 159:1, January 1981, page 71.

Page 154: The drawing on this page is of an object in the collection of the National Institute of Anthropology and History, Mexico. It is used with permission.

Page 163: The diagram is based on "Clams and brachiopods—ships that pass in the night," S. Gould and C. B. Calloway, *Paleobiology*, 6:4, 1980, page 383.

Page 165: The diagram is based on P. G. Williamson's "Palaeontological documentation of speciation in Cenozoic molluscs from Turkana Basin," *Nature*, 293, 1981, page 437.